About Island Press

Since 1984, the nonprofit Island Press has been stimulating, shaping, and communicating the ideas that are essential for solving environmental problems worldwide. With more than 800 titles in print and some 40 new releases each year, we are the nation's leading publisher on environmental issues. We identify innovative thinkers and emerging trends in the environmental field. We work with world-renowned experts and authors to develop cross-disciplinary solutions to environmental challenges.

Island Press designs and implements coordinated book publication campaigns in order to communicate our critical messages in print, in person, and online using the latest technologies, programs, and the media. Our goal: to reach targeted audiences—scientists, policymakers, environmental advocates, the media, and concerned citizens—who can and will take action to protect the plants and animals that enrich our world, the ecosystems we need to survive, the water we drink, and the air we breathe.

Island Press gratefully acknowledges the support of its work by the Agua Fund, Inc., The Margaret A. Cargill Foundation, Betsy and Jesse Fink Foundation, The William and Flora Hewlett Foundation, The Kresge Foundation, The Forrest and Frances Lattner Foundation, The Andrew W. Mellon Foundation, The Curtis and Edith Munson Foundation, The Overbrook Foundation, The David and Lucile Packard Foundation, The Summit Foundation, Trust for Architectural Easements, The Winslow Foundation, and other generous donors.

The Wolf's Tooth

The Wolf's Tooth

Keystone Predators, Trophic Cascades, and Biodiversity

Cristina Eisenberg

ISLANDPRESS
Washington | Covelo | London

Grateful acknowledgment is expressed for permission to reprint the following selection: "The Bloody Sire," copyright 1940 and renewed 1968 by Donnan Jeffers and Garth Jeffers, from SELECTED POETRY OF ROBINSON JEFFERS by Robinson Jeffers. Used by permission of Random House, Inc.

Interior design by Karen Wenk

Library of Congress Cataloging-in-Publication Data

Eisenberg, Cristina.
The wolf's tooth : keystone predators, trophic cascades, and biodiversity / Cristina Eisenberg.
p. cm.
Includes bibliographical references and index.
ISBN-13: 978-1-59726-397-9 (cloth : alk. paper)
ISBN-10: 1-59726-397-4 (cloth : alk. paper)
ISBN-13: 978-1-59726-398-6 (pbk. : alk. paper)
ISBN-10: 1-59726-398-2 (pbk. : alk. paper) 1. Predation (Biology) 2. Predatory animals—Ecology. 3. Food chains (Ecology) I. Title.
QL758.E36 2010
591.5′3—dc22

2009051088

Printed on recycled, acid-free paper

Manufactured in the United States of America
10 9 8 7 6 5 4 3 2 1

Keywords: wolf, wolves, trophic cascades, ecology of fear, Aldo Leopold, predator, prey, food web, keystone, ecology, resilience, adaptive management, apex predator

For my husband, Steve, who gives me hope;
for my father, Zenaido, who taught me about relationships;
and for all the wolves who have crossed my path—long may you run.

What but the wolf's tooth whittled so fine
The fleet limbs of the antelope?
What but fear winged the birds, and hunger
Jeweled with such eyes the great goshawk's head?

—Robinson Jeffers, "The Bloody Sire"

Contents

Acknowledgments

Along the journey of writing this book, I have been supported in countless ways by many people. I am humbled by their collective generosity and insights, without which this book would never have come to be. While it is impossible to thank everyone, I will do my best.

Aldo Leopold, who believed that one must save all the pieces, inspired this work. I am grateful to his eldest daughter, Nina Leopold Bradley, for her steady encouragement from the beginning. My thanks to his younger daughter, Estella, who helped me look at conservation issues from multiple perspectives, and to Curt Meine for shedding light on Aldo Leopold and helping make this book possible. Terry Tempest Williams provided the seed for this work, and for that I will always be indebted to her.

This book began as my master's thesis at Prescott College, which I subsequently augmented with my work as a PhD student in forestry and wildlife at the College of Forestry, Oregon State University, and with additional material on trophic cascades. This book would not have been possible without the exceptional mentorship I have received along the way. I am especially grateful to my ecological mentors, Michael Soulé, James Estes, Robert Paine, and John

Terborgh, for their foundational work in trophic cascades science, and to my academic mentors, William Ripple, Thomas Lowe Fleischner, Hal Salwasser, K. Norman Johnson, Frederick Swanson, Jerry Franklin, Robert Beschta, David Hibbs, Paul Doescher, Martin Vavra, and R. Edward Grumbine, for their guidance and support of my work and this book as it developed. I am thankful for invaluable help from wildlife ecologists Rolf Peterson, Douglas Smith, Diane Boyd, Joel Berger, Kyran Kunkel, and Valerius Geist. I am grateful to John and Charlie Russell, Paul Vahldiek, Doug Dean, and John Rappold for teaching me about living with predators in multiple-use landscapes. My thanks to my mentors in the craft of writing, Rick Bass, Terry Tempest Williams, Allison Hedge Coke, Robert Michael Pyle, Charles Goodrich, and Kathleen Dean Moore. I am grateful to Black Earth Institute founding fellows Patricia Monaghan and Michael McDermott, and to this institution's distinguished fellows and scholars for their inspiration and generous advice on ecological literacy.

I am deeply grateful to Melanie Stidham, Dan Donato, Joe Fontaine, Sandy and Richard Kennedy, Michael McDermott, Trent Seager, Brett Thuma, and my brother Paco for their steadfast friendship and assistance with this book as it developed, which included reviewing multiple drafts. Don Beans, Elizabeth Hughes Bass, Bill Jaynes, and Leigh Schickendantz offered additional friendship and sustenance.

Parks Canada conservation biologists Rob Watt, Cyndi Smith, Barb Johnston, Carita Bergman, and Cliff White helped me learn about relationships at an ecosystem scale, as did their American counterparts and colleagues Jack Potter, Steve Gniadek, John Waller, Tara Carolin, Scott Emmerich, Regi Altop, and Rick McIntyre. My thanks to Kent Laudon of Montana Fish, Wildlife, and Parks and Dr. Mark Johnson for their essential lessons about wolves. Thanks to Roger Creasey, director of the Southwest Alberta Montane Research Program, for enabling my participation in this investigation on how multiple human land uses affect elk, their habitat, and their predators. I thank Carolyn Sime of Montana Fish, Wildlife, and Parks and Greg Hale of Alberta Sustainable Resource Development for facilitating my wolf research. Thanks to Ken Bible of the Wind River Experimental Forest for his hospitality and knowledge.

I thank the many field technicians, volunteers, and writers who joined me afield. I am unable to thank everyone individually because there were so many persons, but I am particularly grateful to Craig DeMars, Dan Hansche, Dave Moskowitz, Neal Wight, and Leah Katz for their expert assistance and to Mark Elbroch for his friendship and advice and for kindly sending me some of his best wildlife trackers. A warm thank-you to Prescott College students and alumni Ashley Burry-Trice, Audrey Clark, Blake Lowrey, and Mark Thorkelson for their hard work and unflagging good spirits under what were at times adverse field conditions. I wish to acknowledge exemplary volunteers Bonnie Sammons, Kathy Ross, and Sandy and Richard Kennedy and their generous contributions to my project over the years. I thank the literary scholars who joined me afield, Annie Finch, Thomas Truelove, and Christina Dickinson, for their reflections on wildness. I wish to acknowledge journalists Douglas Chadwick, Michael Jamison, Olivia Koering, and Brent Steiner, and Karen and Ralf Meyer of Green Fire Productions. I am grateful for their interest in my work via their eloquent writing, photojournalism, and filmmaking efforts.

I am grateful to the institutions that supported me in this endeavor, including the Oregon State University College of Forestry and the Oregon State University Foundation; the Boone and Crockett Club; Parks Canada; the National Park Service; Montana Fish, Wildlife, and Parks; the USDA Forest Service Pacific Northwest Research Station; and the Spring Creek Project. I am particularly grateful to Oregon State University and Prescott College for providing the atmosphere of academic freedom that sustained the creation of this book from beginning to end. All scholars should be so fortunate as to work in such a supportive academic environment. Thanks to The Nature Conservancy for use of its aptly named residence the "Polebridge Palace" to house my field crew when we worked with the North Fork wolves, and to Montana Coffee Traders for fueling this effort. John Frederick and Oliver Meister provided invaluable hospitality and logistical assistance.

I wish to thank the many archivists who assisted me, including those at the Aldo Leopold Foundation; the Bancroft Library at the University of California, Berkeley; Teton Science Schools; the Murie Center; the American Heritage

Center at the University of Wyoming; Oregon State University; the Archives and Oral History Collections at the University of Wisconsin–Madison; and Isle Royale, Glacier, and Waterton Lakes national parks. Thanks to all who allowed me to use their images in this book, particularly Andrea Laliberte for her gorgeous GIS maps.

Extra special thanks to Barbara Dean and Jonathan Cobb at Island Press for their faith in my writing. Barbara's clear vision was instrumental in shaping this manuscript as it developed. She offered encouragement at every stage. Every writer should have an editor like her. Erin Johnson expertly guided me through the many details of bringing this book to life. I am very grateful to Pat Harris for her extraordinary copyediting. Her sharp eyes picked up things I failed to see and helped me hone this book.

This work would not have been possible without my family: my daughters, Bianca and Alana, and my husband, Steve Eisenberg. Their gifts of faith, love, time afield with me, and uninterrupted work time brought this book to fruition.

INTRODUCTION

Visitors from the North

It is the height of summer and I am in northwestern Colorado, on the High Lonesome Ranch, a privately owned property that spans a 300-square-mile swath of the West Slope of the Southern Rocky Mountains. As big as a national park, this working ranch lies northeast of Grand Mesa, just south of the Wyoming border, and contains a good portion of a watershed. Its owners are managing it with conservation and ecological restoration as their primary objectives. Craig DeMars, a skilled ornithologist and the ranch naturalist, is driving us to headquarters to take some guests birding. We bounce along in an old ranch truck at dawn, traveling east on the dirt road that bisects the Middle Dry Fork—a narrow valley palisaded by low, craggy mountains thickly covered with scrub oak and pinyon-juniper forest. This is but one of several valleys that run through the ranch. It rained the night before, which has intensified this arid landscape's colors, leaving it awash in a rich palette of high desert hues—weathered olive sage, yellow-green rabbitbrush, blue-violet mountains.

In the dawn half-light Craig and I make out a herd of elk cows with their calves, foraging in a hay field. We stop and count them. Nearly forty animals. About a mile down the road we see a second, smaller elk herd. As we drive on through this quintessential western landscape, we talk about the research design on this ranch. At the owners' request we are beginning a study of the relationships between elk and deer and their habitat, which includes aspen stands, examining how the removal of predators during the first half of the twentieth century and ranching practices that involved putting too many cattle on a range have affected this relationship. As we consider how we will sample the aspens I see a black animal running across a vivid green alfalfa pasture. I am a conservation biologist who conducts wolf research in the northern Rocky Mountains. Everything about the animal on the High Lonesome Ranch, from the angular set of its blunt, wedge-shaped ears to its fluid, powerful movements, speaks to me of wolf, but my logical mind says no, it can't be. Maybe it's a Black Angus cow. Or a black dog. Craig and I drive along, talking statistics, and I say nothing about the black shape.

Eventually we arrive at headquarters, meet the ranch guests, and take them birding. We stroll companionably below an eroded red rock escarpment, eager to see birds, and soon identify dusky flycatchers and mountain bluebirds in the sage and rabbitbrush. Now and then we hear the mechanical trill of chipping sparrows, but most of the other birds remain silent because of the lateness of the season. Two ravens make lazy circles over russet cliffs to the south. Puddles of last night's rain reflect turquoise fragments of sky. Yellow warblers, their breeding plumage still bright, though it is well past nesting season, and considerably drabber gray plumbeous vireos flit among the tops of the gambel oaks. Fledglings of various species stand out clearly from their sleek elders because of their motley plumage and somewhat out-of-control flight patterns. Amused by the birds, I almost forget the black animal I had seen earlier.

After an hour of birding we go into the ranch dining room for a hearty breakfast and then meet to discuss the ecological research on the ranch. Michael Soulé, who cofounded the science of conservation biology in the 1980s and is the founder and president of Wildlands Network, arrives during the late morning to give a talk about the importance of maintaining habitat connectivity for large carnivores on mixed-use landscapes such as this one—and about how to make this happen on a continental scale. A few minutes later manager Doug

Dean returns from leading a mountain bike tour of the ranch for some other ranch guests. As we are about to sit down to Soulé's talk, Dean takes me aside. Eyes wide with wonder, voice low, he tells me about two pups he saw on the side of the road in the Middle Dry Fork. They had thick forelegs and big feet and acted bold. He describes their husky bodies and gray coloring, and I realize he is not describing coyote pups.

"Visitors from the north?" he asks.

I nod. "Maybe."

It's not exactly safe to be a wolf in Colorado, because of low human tolerance for this species. For weeks we've been referring to the peripatetic canids that may be recolonizing this area from the Greater Yellowstone Ecosystem as "visitors from the north." Indeed, a two-year-old female Yellowstone wolf wearing a state-of-the-art Argos GPS satellite collar made an astonishing 1,000-mile journey to northwestern Colorado and ended up not far from the ranch. SW314F, as she was called, hung around Eagle County, Colorado, for a few weeks but was eventually found dead of unknown causes. Hence our euphemism for wolves on the High Lonesome and Dean's hushed tones in telling me what he's seen. Wolves are listed as a state endangered species, but the Colorado Division of Wildlife has not yet developed a recovery plan.[1]

Dean gives me the pups' location and I realize it is where I saw the wolf earlier in the morning. And all at once everything fits, like puzzle pieces—the wolf tracks and scats that the crew of expert trackers I've assembled to survey wildlife have been finding for weeks now; the black shape coursing fluidly through the alfalfa next to the elk herds. We share our news with Soulé and the ranch owners. As we talk about it briefly, all of us begin to grasp the impact this will have on the ranch's ecology. Wolves touch everything in an ecosystem, from trees to butterflies to songbirds, because of how they influence their prey's behavior and presence, and how that in turn affects the way their prey eat and use a landscape. This has to do with evolutionary relationships that have been in place for millennia, all the ecological pieces present and coexisting, the system working efficiently—until humans removed the apex predators. But this doesn't apply just to wolves; it applies to many other predators—animals such as sharks and sea otters—in other types of ecosystems. We are just beginning to understand the ecological implications of predator removal. Places actively being recolonized by wolves provide vital landscapes in which we can study these relationships.

A Trophic Cascades Approach to Conservation

Ecologists have long understood that predator-prey relationships play an essential role in channeling energy flows within ecological communities. The term *trophic* refers to anything related to the food web, while the poetic term *trophic cascade* refers to the movement of energy through the community food web when predators are removed (or when they return). This dynamic resembles a waterfall and involves top-down regulation of an ecosystem, in which predators have a controlling influence on prey abundance and behavior at the next lower level, and so forth through the food web. Remove a top predator, such as the wolf, and deer grow more abundant and bold, damaging their habitat by consuming vegetation (called *herbivory*) unsustainably. Intensive herbivory can lead to deer literally eating themselves out of house and home and, consequently, to loss of biodiversity and destabilization of ecosystems. Lacking top predators, ecosystems support fewer species because the trees and shrubs that create habitat for these species have been overbrowsed. With top predators in them, they contain richer and more diverse habitat and thus can support a greater number of species such as songbirds and butterflies.

Early wildlife ecologists such as Aldo Leopold advanced ideas about predation's role in maintaining ecosystem structure, composition, and function. Although it would be decades before the term *biodiversity* would be coined, Leopold's observations foreshadowed many of the concepts that continue to shape modern conservation.[2] Since his death in 1949 the new science of conservation biology, which focuses on sustaining biological diversity, is helping us move toward more responsible and better-informed management of our natural resources. Some agencies have adopted an ecosystem management approach, as opposed to a single-species approach, taking into account the full spectrum of diversity and the ecological processes that sustain it.

In this book we will explore the science and conservation implications of trophic cascades. To do so we will take a wide view, one in which we will dig deeply into the disciplines of community ecology, landscape ecology, wildlife biology, and conservation biology. This interdisciplinary approach will help us gain a clear understanding of what trophic cascades are and why top predators matter in sustaining the optimal functioning of ecosystems.

We will begin by looking at trophic cascades within the context of the web of life. I will share accounts from my fieldwork that illustrate the immediacy of foundational concepts such as the ecology of fear and resource selection. We will explore an evolutionary perspective on trophic cascades by examining how late Quaternary megafaunal extinctions left a deep imprint on the landscape patterns and predator-prey interactions we observe today. Because awareness of trophic cascades is important as we address global environmental changes, we will review the promising body of research that highlights this connection. I provide a glossary as a reference to help illuminate some of the key terms used by the researchers doing this work. By expanding the depth and breadth of science across many different environments, these studies are changing our perception of how nature works and bringing conservation goals within reach.

Ecosystem restoration provides some of our best hope for the future. Accordingly, we will explore how resource managers can use knowledge of trophic cascades to guide recovery efforts on public and private lands. This science can be used to move forward the bold and at times controversial vision of rewilding the North American continent. This vision calls for slowing today's rapid extinction by reconnecting and restoring habitat and wildlands, with focal strategies that include conservation of large carnivores and the ensuing recovery of mechanisms such as trophic cascades. Trophic cascades science provides essential and practical knowledge that fits within the rubric of ecosystem management. Here I provide my own recommendations for local and landscape-scale workable applications of what we are learning about interactive food webs.

Trophic cascades are an ecosystem's stories writ large upon aquatic and terrestrial landscapes. "Visitors from the north," wolves and other top predators, leave distinct patterns—easily observable effects such as a flush of aspen growth or luxuriant kelp forests—in formerly impoverished systems. To those of us studying these dynamics, the landscapes in which they occur are landscapes of hope. Ecosystems speak to all of us—researchers, managers, students, and plain members of the biotic community. If we pay close attention they will tell us what to do as we strive to heal the ecological wounds caused by human impacts and to address global change.

PART ONE

Web of Life

CHAPTER ONE

Patterns in an Ecosystem

We bushwhacked through an old burn at first light in a cold September rain mixed with snow, slipping on the blackened bones of downed lodgepole pines. It had been three years since I'd come this way, past a curving reach of the Flathead River and the old ranger station, through a locked gate, and into a vast, fecund meadow a few miles south of the US-Canada border in Glacier National Park. As steeped in ecological history as Yellowstone National Park's famed Lamar Valley, but far less known, the meadow lay beyond the burn, although my field crew and I couldn't see it yet. It held ecological stories plainly told as patterns in an ecosystem, which we were there to record.

When wolves (*Canis lupus*) recolonized northwestern Montana in the 1980s they chose Johnson Meadow, a secluded opening in a lodgepole sea, as their first home. In 1986 renowned wolf biologist Diane Boyd, then a graduate student, confirmed the first denning activity here after a sixty-year human-imposed wolf absence.[1] Glacier National Park administrators keep this place closed to the public but occasionally allow researchers in—and then only when the resident Dutch pack, which is radio-collared, travels away from the den. Trouble is, the wolves seldom leave, lingering at the den site until long past spring whelping season, feeding on the abundant deer (*Odocoileus* spp.) and elk (*Cervus elaphus*) with which they share the meadow. When they do leave they tend to travel one or two miles from the den, remaining in the general vicinity to hunt or rest with their pups at areas called rendezvous sites.

I had last visited this den on a benign autumn day when the aspens blazed like souls on fire against a deep blue sky and thistledown floated on the wind. I had been helping with a study of how wolves select their den sites. Now I had returned to conduct research of my own: a study of trophic cascades involving wolves, elk, and aspens (*Populus tremuloides*) in the Crown of the Continent Ecosystem.[2] This ecosystem spans the US-Canada border, one of two in the lower forty-eight coterminous states that contain all species present at the time of the Lewis and Clark expedition.

I had chosen to focus my research on the aspen because, although it is the most widely distributed tree species in North America, it has been declining in large portions of the intermountain West since the 1920s. Aspens reproduce clonally, sprouting from extensive root systems, and provide critical habitat for diverse species of wildlife and plants. They offer the richest songbird habitat, second only to the interfaces between streams and land, called *riparian* zones. Because aspens can support such profligate biodiversity, their decline has created pressing research and conservation needs. I had chosen to study elk because their impacts on aspens are greater than those of other hooved animals (called *ungulates*).[3] The steepest aspen declines have occurred in areas of elk winter range, linked to predator removal and influenced by disease and climate variability.

Trophic cascades refers to the relationships among members of a biotic community: predators, prey, and vegetation. In 1980 marine ecologist Robert Paine coined this elegant term to describe this interaction web.[4] These cascading, predator-driven, top-down effects have been reported in all sorts of ecosystems, from the Bering Sea to rocky shores to montane meadows. As in all of these systems, the fundamental three-level food web I studied indirectly touched many other members of the biotic community, which in this case included songbirds.

Rooted in flesh-and-blood encounters between predator and prey, trophic cascades involve passage of energy and matter from one species to another. Each act of predation subsumes one life so another can continue. Predation can have strong direct and indirect effects in food webs, making nutrients such as nitrogen flow through ecosystems, with significant consequences for community ecology.[5] Wildlife corridors, such as the one I was working in, are characterized by heightened species interactions and nutrient flow. They provide natural laboratories where ecologists can learn much about trophic cascades. In 1935 preeminent American wildlife ecologist Aldo Leopold noted how predators help increase species richness and how their presence affects everything from prey to plant communities. He eloquently wrote about these relationships and the lessons he had learned from them about ethical resource management in his book *A Sand County Almanac.* Current trophic cascades research is adding to our awareness of these relationships. My time in Johnson Meadow was part of my effort to elucidate these dynamics.

A few feet into the lodgepole jackstraw we came upon the first wolf scat—two inches in diameter, oxidized white, filled with ungulate hair, a bold territorial marker left in a well-worn path. Generations of wolves circling the meadow and then arrowing into it had made this path as their tracks homed into the den area. The trail sped our passage through the old burn, our feet finding easier purchase where so many wolves had trod. The burn stopped abruptly at the meadow, which remained wet and marshy in some spots year-round and thus had been singed only lightly. We soon entered a network of other wolf trails that wove through tussocks of tassel-topped fescue and fireweed gone to seed, taking us deeper into the meadow.

Covering approximately ten square miles, Johnson Meadow held five large aspen stands and a long-abandoned homestead, now little more than a few boards weathered silver and a midden heap in a damp declivity. Low-lying glacier-smoothed mountains rimmed the meadow. Anaconda Peak's rocky southern face rose sharply above a series of soft green ridgelines that faded into the north. To the south Huckleberry Mountain's rounded bulk breached a fog bank, its shoulders a mosaic of burned and unburned patches. It had provided first-rate grizzly habitat until a recent fire took out much of the timber and berries.[6] The meadow curled east, revealing its full expanse and secrets gradually. And indeed, given that few humans were allowed to enter and it was completely hidden from the road, it felt like a secret meadow. It seemed no less primordial and wild than I recalled. Bones everywhere: deer, elk, and moose. Strategically situated lays, places where the pack had rested and perhaps surveyed the landscape, matted the tawny grass. The wolf sign intensified the farther we ventured.

The wolf trails all led into a large lay. My daughter Alana, in the field with me that day, photographed a bull elk skull. It had been there so long its cranial sutures had gone mossy and the scarlet leaves of a frost-singed geranium had sprung through an eye socket. Later I would be struck by the lyrical beauty and symbolism of that image—the juxtaposition of life and death it contained and which coursed through the meadow like a leitmotif. This juxtaposition had shaped trophic cascades ideas from the beginning.

People have been noticing trophic cascades for centuries. As early as 2,500 years ago the Chinese noted cascading top-down effects and promoted the use of predators to lessen crop damage. In their orchards Chinese farmers established nests of predatory ants to reduce numbers of caterpillars and large boring beetles.[7] In 1859 in *The Origin of Species*, Charles Darwin documented an interactive food web, comparing the height of fir trees in fields grazed by cattle with those of firs in fields without cattle. He found that grazing cattle completely prevented forests from becoming established. In another case study he linked the relationship between predator presence (domestic cats), lower prey (mice) populations, and an increase in bee populations (whose hives the mice plundered).[8]

Early British ecologist Charles Elton noticed these patterns, and they inspired his groundbreaking book *Animal Ecology*, first published in 1927. In it he described ecology as "scientific natural history" and presented concepts we have come to know as its basic tenets: food chains, the role of size in ecological interactions, niches, and the food pyramid. Influenced by Darwin, he depicted nature as an integrated economy in which food web members exchange energy.[9] Some managers of that era, who were grappling with disease outbreaks, species invasions, unstable game populations, and degraded ranges, welcomed this community ecology perspective.

Elton conceptualized the food chain with predators at the top, followed by herbivores, and vegetation and simpler organisms at the bottom. As food chains crossed and connected in various ways they formed complex food webs, which could be seen as a map of trophic activity. In arctic systems the web would be simple; in the species-rich forests of the humid tropics the web would be vastly more complex. Today ecology continues to be built on Elton's model and on the exchange of energy across trophic levels.

Aldo Leopold, who was a close friend of Elton, observed how these ideas played out in the real world. He documented widespread sharp increases in North American populations of ungulates from the 1920s through the 1940s. The first to apply the term *irruption* to this phenomenon,[10] he identified the irruption sequence as removal of a top predator, then release from predation of its herbivore prey, which leads to an increase in prey numbers, followed by overbrowsing and overgrazing—what today we call a three-part trophic cascade. He also noted cascading trophic interactions driven by wolves in Mexico's Sierra Madre Occidental, contrasting that with the extensive deer and elk irruptions throughout North America in areas where wolves had been removed, such as Arizona's Kaibab Plateau. In Mexico the intact flora, fauna, and watershed he explored harbored an abundant, but not excessive, deer herd thriving among abundant populations of all species of large carnivores present historically (e.g., grizzly bears, *Ursus arctos*; cougars, *Puma concolor*; and wolves).[11] In his hunting journal he expressed astonishment that this was the first time he had observed land that wasn't sick, what he referred to as a "biota in its aboriginal health." But beyond recognizing food web dynamics and their significance, he identified ecologists' responsibility to utilize these emerging ideas to help solve resource management problems. He advised his wildlife students at the University of Wisconsin, "To keep every cog and wheel is the first precaution of intelligent tinkering."[12]

Leopold's observations contributed to a debate igniting in the scientific community in the mid-1940s about the relative role of predation in shaping communities. In 1960 a landmark paper by ecologists Nelson G. Hairston, Frederick E. Smith, and Lawrence B. Slobodkin, collectively known as HSS, created a flash point. They argued that the world is green because predators limit herbivore populations (top-down control).[13] At the time they advanced what became known as the green world hypothesis, the scientific community commonly accepted that vegetation-driven (bottom-up) processes were the primary forces structuring populations. Since the HSS paper was published, researchers have been looking for and measuring patterns of resource use by predators and their prey. The earliest trophic cascades research was fueled by excitement and synecdoche: meticulous observations and experiments that held bigger truths.

The debate continues, exacerbated by these interactions' tangled architecture. Today most scientists agree that trophic cascades occur. According to marine ecologist James Estes, the argument now focuses more on *where* and *why* they occur.[14] For example, Yellowstone researchers William Ripple, Robert Beschta, and others suggest that since the reintroduction of wolves in the park in the mid-1990s, cascading effects have included a reduction in the elk population, changes in herbivory patterns, a reduction of mesopredators (e.g., coyotes, *Canis latrans*), and recovery of woody browse species. Wolves recolonized Banff National Park on their own. Here Mark Hebblewhite and colleagues found the above associations and an increase in biodiversity, which included songbirds and beavers (*Castor canadensis*).[15] However, in the similarly recolonized upper Midwest these effects are not widely evident, other than the trophic cascades observed in Isle Royale National Park by Brian McLaren and Rolf Peterson, possibly due to extensive human modification of this ecoregion.[16] These examples illustrate that because ecosystems are highly dynamic and complex, generalizations about trophic cascades are difficult, making their application to specific management issues more challenging.

Ecologist Mary Power is well known for her studies of freshwater trophic cascades. In her 1992 landmark synthesis of top-down and bottom-up forces in food webs she argued that while top-down effects often prevail, one must consider bottom-level (plant) productivity and its potential for dynamic feedback effects on adjacent and nonadjacent (herbivores and predators) trophic levels. Power found that both top-down and bottom-up forces prevail in different settings and urged researchers to consider how increasing plant productivity in all types of systems might affect predator-prey interactions. She noted that disturbances often make ecosystems more productive and thus can further alter these relationships.[17]

Since the 1990s ecologist Oswald Schmitz has been investigating trophic cascades involving grasshoppers and spiders. One of his objectives has been to determine whether top-down or bottom-up forces dominate. By manipulating predator presence and other variables, he found both forces present, with the former the strongest. He discovered that regardless of the strength of trophic cascades, they have important effects on plant community composition, with deep implications for biodiversity and ecosystem function.[18]

In aquatic systems, perhaps the best-known and most dramatic example of a trophic cascade involves Estes' work in the Aleutian Islands, where he and colleagues investigated the recovery of sea otters (*Enhydra lutris*), the associated reduction in their sea urchin (*Strongylocentrotus* spp.) prey, and subsequent recovery of kelp forests (*Laminaria* spp. and *Agarum cribrosum*). Because kelp forests are highly productive components of rocky marine coastlines and support diverse communities, this otter-driven cascade has high conservation relevance.[19] A more complex, multilayered food web occurs in South Pacific coral reefs. Here fishing by humans has been linked with a decrease in predaceous fish, an upsurge in the sea urchin population, and a reduction in biodiversity. But these relationships are not as linear as in other systems. Consequently what occurs here can be thought of as a "trophic trickle"—what happens when a three-level trophic cascade's potentially strong effects become diluted across multiple intertwined levels.[20]

Trophic interactions such as these tell a story made all the more compelling by the breadth of our still incomplete understanding of community dynamics. It's a story that bears telling for the lessons it can teach us about sustaining the richness of life in all its forms. Even as we strive to expand our knowledge of species interactions, biodiversity is decreasing at an alarming rate. One-fifth of all the bird species in the world have gone extinct in the past 2,000 years, and 11 percent of the remaining 9,040 known species of birds are endangered. The number of plant and animal species on the earth is declining at a rate 1,000 to 10,000 times higher than in prehistoric times. Extinction estimates vary, depending on researchers' methods, with most predicting a loss of approximately 7 percent of species globally per decade. Biodiversity loss has become a crucial issue since the late 1990s as human-caused ecosystem modifications continue to precipitate extinction. Trophic cascades increase plant growth; the resulting energy surge drives nutrients across ecosystems, with significant effects that include an increase in biodiversity.[21]

Like all aquatic and terrestrial landscapes, Johnson Meadow is a palimpsest written over for millennia by its wildlife and human inhabitants. Landscapes speak to us through these marks. Trophic cascades research involves interpreting historical marks as well as current wildlife movements and trends. To the north of the lay where the wolf trails converged in the meadow stood an aspen stand atop a low knoll. It contained a smattering of conifers. As we entered this leafy glen on a wolf trail, we found a cache of purloined toys that suggested the den was nearby: plastic soft drink bottles pocked by pups' sharp teeth, frayed strips of fiberglass insulation, and a thoroughly gnawed road sign. A shrub and sapling thicket rendered the trail nearly impassable to humans and hid the den from view until we stood directly in front of it.

Relatively unchanged since I'd last seen it, the den nestled into the root-ball of an ample spruce. On the bare earth around it we found wolf hair and pup scat, bones, ungulate hair, and several wolf beds. It had three wide openings, set flush into the ground, and was rather large, as wolf dens go, because it had been used for many consecutive years. A faint but unmistakable fetor clung to it. Just inside the shallowest opening I found the culprit: a dead juvenile spotted skunk. The bedraggled creature bore a few chew marks and a small pile of pup scat tellingly deposited on its back.

Interesting as I found these artifacts, I hadn't come to Johnson Meadow to study them or the wolves that had left them. I was here to measure aspens—for these trees told a powerful ecological story about how the wolves' presence affected their prey's behavior and density, and how that affected the way aspens and other vegetation grew. And there were few better places to see how these relationships unfolded, because in addition to being the inner sanctum of wolf activity in the park, Johnson Meadow is considered by managers to be prime elk winter range, where herds gather when snow blankets the surrounding mountains.

Elk initially survive the fallow season by nosing through the snow to feed on dried summer grass, which has a lower nutritional value in winter. As the snow deepens they shift their feeding strategy to woody browse species, including aspens, cottonwoods (*Populus* sp.), and willows (*Salix* sp.). They must consume considerable amounts of twigs and bark to meet their nutritional needs for protein and minerals.[22]

The trees around the den grew vigorously, with many saplings reaching beyond browse height—the height an elk can reach to eat, usually six and a half feet. However, these trees also showed evidence of moderate elk herbivory. Johnson Meadow provided a good example of how wolf presence doesn't stop ungulate use of an area but has a strong bearing on how the ungulates use resources. Upon closer inspection the browse pattern here suggested that rather than standing around eating aspen sprouts to the ground, the elk had exercised restraint, perhaps taking one bite or two, looking up to scan for wolves, moving on, and then stopping to browse some more.

Defined in some systems by fear more than by the actual act of predation, trophic cascades mean that the presence of a predator, such as the wolf, affects prey foraging behavior as prey try to find an optimal balance between fear of being eaten and meeting their metabolic needs. Wildlife ecologist Joel Brown and colleagues termed this phenomenon *the ecology of fear*.[23] The ineluctable play of energies involved in fear of predation is no less real than the act of predation. This predator-prey death dance, evinced by the bleached bones strewn about the meadow, the braid of wolves' voices at dusk, and ravens on the wing lured by the coppery scent of fresh blood, causes human and wild animal hearts to beat faster and cascading effects to ripple throughout food webs. The immediacy and mystery of these relationships drew me as a scientist and compelled me to consider questions ecologists and naturalists had been pondering for centuries.

Cycling predator and prey populations leave marks. In terrestrial ecosystems with large carnivores, one of the most blatant is the recruitment gap. When wolves abound, fear of predation causes elk to stay on the move and be more vigilant. They spend less time feeding, which can enable browse species to persist until they are above browse height. With wolves absent, elk become less vigilant and more abundant, browsing freely and heavily, not permitting young plants to grow into adult trees.

Conservation biologist John Terborgh and colleagues tested what happens to systems that harbor herbivores unchecked by predation. In 1986 Venezuela's Lago Guri flooded the surrounding hilly terrain, creating hundreds of land-bridge islands. Some of the islands were too small to sustain vertebrate predators. This event facilitated a predator removal study that enabled observation of whole ecosystem responses. On islands lacking predators Terborgh found increased populations of generalist herbivores, such as leaf-cutter ants (*Atta* spp., *Acromyrmex* spp.) and red howler monkeys (*Alouatta seniculus*), and ensuing destruction of the vegetation, so dramatic as to be easily observed by a casual visitor. Islands sufficiently large to support predators appeared lush and green, with abundant leaf litter and saplings, while predator-free islands had a denuded forest understory. This trophic cascade vividly illustrates the stabilizing role of predators. We will explore this story further, as well as Terborgh's other trophic cascades research, later in this book. Perhaps the most urgent value of his work lies in what it can teach us about maintaining biodiversity.[24]

The herbivory pattern shaped by wolf extirpation had left an equally bold signature on Johnson Meadow's aspens. Tree coring and ring counts would suggest that the new flush of aspen growth I was observing dated back to the mid-1980s, when wolves returned. However, trophic interactions get complicated in speciose systems (those that contain many species).[25] The challenge before me lay in measuring their strength—determining what in this system directs the architecture and flow of trophic cascades. In addition to sampling vegetation, I would be using data from radio-collared elk and wolves to study these relationships spatially. These data would give me a window into the habitat selection, movements, and feeding habits of each tagged animal and a greater awareness of how animal behavior fits into trophic cascades hypotheses.

My field crew and I began putting belt transects (lines used to study ecological conditions) into the aspens to measure the plant community. Not surprisingly, the trees all told the same tale—of fear and vigilance, of the predator-prey dance. I planned to sample every stand. The wolves kept their distance, allowing us to work. They stayed out of sight, but we knew they were nearby because we heard their howls and picked up their radio-collar signals' strong pings. And so we went about our business, moving quickly, fingers numbed by the freezing rain, which mostly didn't let up. More rain. More transects, with wolves howling in the background.

On our final day in the meadow we had the good fortune to be joined by Kyran Kunkel, a wildlife biologist who in the 1990s conducted landmark research on predator-prey interactions in Glacier National Park's North Fork.[26] I had one more aspen stand to survey. It graced the southern tip of the meadow, about a mile from the den, an area we had yet to explore. As we walked over, the rain finally stopped. The sun broke through the clouds and made the wet grasses glisten. I unzipped my rain parka. Someone started singing, our group of five in fine spirits. All at once, out from a fir copse in front of us ambled a grizzly—impossibly huge, the color of dark chocolate. The wind ruffled the long hairs on its muscular hump. No one moved. We silently watched the bear. It nosed the earth and then began lissomely pawing up great hunks of it, searching for small, ground-dwelling mammals and popping them whole into its mouth when it found them, its heavy jaws making quick work of them.

The bear methodically dug out burrow after burrow. Eventually it paused in its labors, lifted its snout, tested the air, and caught our scent. Kyran and I exchanged glances as we waited to see what it would do. The bear began to pace, moving its head from side to side until it found us. I glassed it with binoculars and noted its long claws and the power it radiated with every gesture. Its small eyes seemed to make contact with mine through the binoculars, though I knew that was impossible. All the same, the fine hairs on the back of my neck stood up. Unlike most bears facing humans, this one stood its ground. It told us emphatically with its body language that it would not vouchsafe us entry into the southern end of the meadow.

Three hundred feet away, give or take. Safe, but not safe enough. Not when you're dealing with a hyperphagic bear—one liable to become aggressive if interrupted during its prehibernation feeding frenzy. Not when you're dealing with an animal easily capable of sprinting at speeds in excess of thirty-five miles per hour. We followed yet another wolf trail to the top of a knoll that offered the highest vantage point in the meadow, to watch from a longer distance. Standing under a lone, ancient ponderosa pine, we waited. The bear resumed feeding. It gave us looks now and then to let us know it hadn't forgotten us.

Late morning ran to afternoon and nothing had changed. The bear held us at bay, giving us an immediate awareness of our own place in the food web. A murder of ravens scythed through the air, black arrows on a western trajectory toward the Flathead River. We surmised that there must be a carcass over there. A northern harrier wheeled low circles below us as it hunted, its pearly butt-patch gleaming. We sat in a sward where wolves had once bedded, judging from the flattened grass and tufts of lupine hair, and discussed ecology, speaking in measured tones, listening and telling in turns. We imagined aloud what it must be like to be one of the wolves that had lain in that spot, to see the meadow through a wolf's eyes. And in such imaginings we gained a landscape-scale perspective we had not had before. Mindful of the bear's movements, we talked about our imperfect understanding of the relationships in a food web—what happens when two or more species of predator share an area—how this affects prey behavior and everything else and how this shapes trophic cascades. How the relative strength of trophic cascades can be controversial, given their complexity. We talked about the priceless value of transboundary wildlife corridors that cross international borders, such as the one in which we found ourselves, one of the few places in North America where all native carnivores persist and have abundant prey and adequate habitat. And we marveled at the wholeness of this place—a wholeness essential to our well-being as a species.

That bear never did let us pass. We left at midafternoon without putting transects into the last stand, my data set incomplete. But paradoxically, that encounter felt like the perfect closure to my time in the meadow because of the way it stopped me in my tracks at the end of a busy field season and urged me to pay attention to where I stood on the trophic cascades trail. Stepping aside from that bear was not an act of submission but an act of admission that other members of a food web can and will take precedence when we meet them on their own terms. Weeks later, while analyzing data, I opened my aluminum field clipboard and the ripe, heady scent of September in Johnson Meadow came spilling out, along with fine bits of golden leaf litter and twigs—the detritus of my work. And the memory of that last day gave my data immediacy.

On the day after the bear encounter—the autumnal equinox—I drove to Alberta, where my study sites ranged northward along the Rocky Mountain Front. On the way I stopped on the eastern side of Glacier National Park, in an aspen parkland named Two Dog Flats by the Blackfeet Indians, to retrieve an increment borer—a tool I had been using to core trees. It consisted of a hollow, two-foot-long and half-inch-thick enameled cylinder that held a corkscrew-pointed steel shaft. I had accidentally dropped it while running from a rapidly advancing thunderstorm the previous month. When the storm struck I had been putting in transects, finding a far different pattern here than in Johnson Meadow.

I began to walk up a broad, dun-colored prairie patch that swept gracefully up to the base of a thousand-foot massif of crumbling rosy argillite. Aspens ringed this elk wintering ground, the stands arranged in the long streamers characteristic of aspen parklands east of the Continental Divide. The air held a thin scrim of smoke from a midsummer fire, which still smoldered on the southern side of St. Mary Lake. The wind carried the distant, ragged bugle of a bull elk—a primal reminder of the turning season. Soon the snow and cold wind would come, and these trees would once again provide both shelter and winter food for masses of elk. But for how many more decades was uncertain.

This place differed dramatically from Johnson Meadow in two fundamental ways: it had no sustainable wolf population, a result of human pressure, and there was virtually no aspen recruitment. Mostly wolfless since the 1920s, Two Dog Flats aspens had been browsed relentlessly by elk. Wolves had been trying to recolonize this area since the 1980s, without much luck until recently, when a pack denned and produced pups nearby, on the Blackfeet timber reserve. However, the ranchlands along the Rocky Mountain Front just outside the park did not provide good wolf habitat; a significant portion of that pack had been removed because of livestock depredation. And although this area supported abundant cougar and grizzly populations, the presence of these predators did not deter elk herbivory on aspens in the northern Rocky Mountains.[27]

The plant productivity was similar to that in Johnson Meadow, the aspens sending up several hundred sprouts in each transect; however, the similarity ended there. These sprouts' zigzagging trunks and scarred stems showed that elk were hitting them hard. Many had become stunted from overbrowsing, growing in thickness but not in height. They faced possible death from unchecked herbivory. The only new stems growing above browse height occurred amid refugia created by hawthorn thickets armed by vicious three-inch spikes. With the exception of these sheltered stems, those that had lasted beyond a couple of seasons had blackened trunks, heavily scarred from decades of elk gnawing. The aspen stands in Two Dog Flats were not unique; similar dynamics have been reported in other elk winter ranges. Even-aged, single-storied stands grow throughout North America, with most aspens older than eighty years, no middle-aged stems, and few stems recruiting in areas with elk during years when predators are absent.[28]

Three-quarters of the way upslope I found the increment borer, its orange enamel bright against the faded grass. I retrieved it and drove on up to Alberta for some research meetings. Just north of the border I stopped in the Belly River Valley, not far from a campground in Waterton Lakes National Park. Aspens lined this shoestring valley, their gilded foliage in crisp contrast to the inky conifers that surrounded them. I walked a good portion of that vale, noting aspen age classes and growth patterns. Along my path I found a lively tattoo of wolf tracks.

Wolves recolonized the Belly River country in the mid-1990s, their denning confirmed by wildlife biologists in 1994 and multiple times since then.[29] This remote valley also contained elk winter range. And the trees here had prospered, with thousands of young aspens growing straight, white trunked, and slender. They had rapidly surpassed browse height, with only traces of elk herbivory. Stories on the land. It was too late in the season for more vegetation sampling, so I would have to wait until the next summer to put in transects. I took some photographs and drove on to Pincher Creek to meet with resource managers and co-researchers, trophic cascades very much on my mind.

Trophic cascades are directly related to the keystone species approach to conservation. In 1969 Robert Paine created this term to apply to sea stars (*Pisaster ochraceus*), which functioned as top predators in the rocky intertidal zone he studied. His concept is fundamentally about species diversity—that the presence or absence of one key species influences the distribution and abundance of a great many others. This metaphorical term, which refers to species that function similarly to the keystone of an arch (remove the keystone and the arch falls apart), was intended to convey a sense of the unexpected consequences of species removal. While it has suffered from overly broad application, the underlying concept remains essential because it helps us understand how species loss will affect ecosystems. It emphasizes an integrated approach to conservation and to understanding how ecosystems work. The effects of keystone species, also referred to as *strongly interactive species*, include habitat enrichment, symbiosis, and competition—the sort of dynamics at the heart of trophic cascades.[30] Science provides our best tool for probing what these interactions mean to ecosystem function and structure. Ecologists working in a variety of systems are showing how loss of a keystone predator causes disruption of ecological processes that can lead to simplification of biological communities. These processes include predator-prey interactions, nutrient cycling, and seed dispersal. Research can help us recognize these losses and thereby act to prevent others.

Whether predators control their prey and, if so, how these interactions may be shaping ecosystems have been leading questions in ecology since 1960, when HSS created their controversial green world hypothesis. While these questions are far from simple, in subsequent chapters we will further explore this hypothesis and see how landscapes tell stories and how researchers are finding remarkably similar patterns in ecosystems worldwide. In this time of rapid global change, trophic cascades science is helping us understand how nature works. And the answers we are finding can help us sustain biodiversity in this world.[31]

CHAPTER TWO

Living in a Landscape of Fear: Trophic Cascades Mechanisms

A doe burst out of the forest and tore across the meadow, two wolves in close pursuit. This drama unfolded not twenty feet from where my young daughters and I knelt in our garden peacefully pulling weeds, our pant legs wet with morning dew. One black, the other gray, the black wolf in the lead, they closed in on the doe's haunches. In less than two heartbeats they pierced the deep wood on the far side of the meadow, leaving a wake of quaking vegetation.

We live at the base of a mountain in northwestern Montana. As wild as the wildest places in the lower forty-eight United States, it isn't quite paradise, although the handful of us who live here think it comes close. Midway up the mountains, overgrown clear-cuts show up as yellow-green rectangles against the darker green of old-growth forest. From our cabin you can walk due east beyond the state forest lands and not encounter much more than federally protected wilderness for a hundred miles.

Landscapes shape us and speak to us on a primal level. Most of us have a landscape we intuitively comprehend. This is mine. I open the front door of my cabin and find wolf tracks pressed into the snow. In spring, even before I see the grizzly lumber out of the forest to dig roots, I smell its ripe essence. These discoveries give me pleasure and an unspoken awareness of the natural order of things.

Humans also have a primal relationship with large predators. This relationship has been eloquently elucidated across the ages in Paleolithic petroglyphs of dire wolves and other creatures sharp of tooth and claw and in medieval paintings of wolves menacing sheep. Wolves began to recolonize our area in the early 1990s. Since then we had been hearing them howl from the shoulder of our mountain and occasionally finding their tracks. But we had never seen them—not until that misty August morning when they ran across our meadow. For some long moments after they passed we knelt motionless in the garden, at a loss for words. Then curiosity kicked in and we stepped outside our small fenced yard to follow the wolves' trail.

I marked one track, and from it we located others laid out in a gallop pattern. We even found the spot where one of the wolves had turned to look at us, a motion that had caused its left front foot to break forward. Fascinated, we continued to follow the subtleties of their trail—which sometimes consisted of little more than a few bent blades of grass that even as we watched were springing back upright. And I wondered how many other times wolves had run through our land and I'd missed the evidence.

In the fifteen years since wolves returned, the deer had been behaving differently—more wary, not standing in one place, eating all the shrubs down to nothing. After the first three years I seldom saw deer browsing in the meadow, and then only for brief periods. And after a decade the meadow was nearly gone, with shrubs and young aspens filling in what used to be open grass. Until we saw the wolves hunting, I had never actually observed a trophic cascade in action.

The Green World Hypothesis

In 1969 Loren Eiseley wrote an evocative story, "The Star Thrower," about a man who walked the strip of wet sand that marks the tide's ebb and flow, tossing sea stars that had washed ashore back into the ocean. Motivated by a need to save them from death, each day he returned the stars to the ocean. At one point in the narrative, Eiseley commented on the apparent futility of this task.

In the real world the star thrower is a scientist, and death is running even more fleet than he across every ecosystem on this earth. Like Eiseley's star thrower, Robert Paine is motivated by promoting life, although the results of his actions are far less futile and best described as utterly Eltonian. Considered the father of trophic cascades science, Paine has spent most of his life studying aquatic communities. I visited his University of Washington lab, which overflows with the products of a long and illustrious career: books and papers he's authored, photographs of the intertidal world he's studied for so many years. He sums up his work in two sentences: "You can change the nature of the world pretty simply. All you do is get rid of one species."[1] As a young scientist he proceeded to do just that, experimentally removing sea stars in control plots, thereby creating profound changes in the intertidal community he studied.

In chapter 1 we were introduced to the trophic cascades concept created by Nelson Hairston, Frederick Smith, and Lawrence Slobodkin (HSS). In 1960 they proposed that vegetation patterns are determined primarily by patterns of food consumption by herbivores. They further suggested that the act of predation shaped herbivory. Therefore, herbivores unchecked by predators would have a great influence on vegetation. Known as the green world hypothesis, their ideas provided an ingeniously simple explanation for why the earth is green.[2] This hypothesis resulted from a seminar at the University of Michigan, in which scientists met to discuss central issues in terrestrial ecology and arrived at this remarkable conclusion. Their ideas were inspired by Charles Elton's food pyramid.

As a new professor at the University of Washington, Paine tested the green world hypothesis. The idea for a simple experiment that essentially involved playing God in the rocky intertidal zone came to him during a visit to the Scripps Institution Wharf in California, as he stood watching the carnivorous sea star *Pisaster ochraceus* devour the mussel *Mytilus californianus*. What if he removed sea stars to see what would happen? The next year, National Science Foundation funding in hand, he settled into the University of Washington and began his research. Month after month he traveled to Mukkaw Bay, at the northern tip of the Olympic Peninsula. There, on a rocky crescent of shore, he hurled sea stars into the ocean. In his control plot he did nothing. As he continued removing stars, the assemblage of species on the rocks gradually began to change. Within one year the sinister implications of his experiment became all too graphically obvious. Where *Pisaster* flourished, so did the vegetation. Where *Pisaster* had been removed, the mussels took over, crowding out other species and eating all the vegetation, until little more than a dark carpet of mussels and barnacles remained. The paper Paine produced for the *American Naturalist* about this research turned out to be one of the most influential journal articles in the history of ecology. It provided one of the first examples of an ecosystem that had been transformed by trophic cascades.[3]

Mentored by Frederick Smith of HSS fame, Paine was also influenced by Charles Elton and Charles Darwin. He thinks the green world hypothesis may have had its genesis during Smith's 1957 course at the University of Michigan on the natural history of freshwater invertebrates, on the kind of spring day when professors don't really feel like teaching and students don't really feel like sitting in class. That day Smith looked out the window of the zoology building, which faced a courtyard, and said to his students, "We're not going to have a field trip today. I want to teach you how to think instead. Tell me, what do you see out there?"

"A tree," said some bright person.

"What color is the tree?" asked Smith.

"Well, it's got green leaves," said another student.

"And why aren't the grazers out there—the insects—eating all those leaves? What makes the leaves green?" asked Smith.

"Well, it's photosynthetic pigment," said someone.

That wasn't quite what Smith was after; he wanted them to explore ecological relationships between organisms. And so the conversation continued, back and forth. For Smith had been thinking about these ideas for quite some time: he was a deep thinker. So he looked out the window, and this argument of immense consequences followed linearly from that. What came to be known as the green world hypothesis was first articulated on that bright spring day in Smith's zoology class, and it fueled the fires of enthusiasm in the graduate students at the University of Michigan and rocked science.[4]

The Architecture of a Trophic Cascade: Predator-Herbivore-Vegetation

When HSS formulated their hypothesis, the scientific community commonly accepted that bottom-up processes, mostly related to competition between species, were the primary forces shaping populations. The green world hypothesis provided an alternative view of population regulation driven by top predators, via carnivory. It enabled predation and grazing to play roles equally important to resources or habitat, with predation having the key role for some populations and resources having the dominant role in others.[5]

HSS organized the food chain into three trophic levels. At the lowest trophic level they put producers (plants), with consumers (herbivores) at the next higher trophic level and predators at the top—as in Elton's food pyramid. Resources limit each trophic level. HSS noted interspecific competition among members of each level and concluded that because herbivores are seldom food limited, they appear to be most often predator limited and thus unlikely to compete for common resources.[6] They offered the example of the vast carbon deposits that had accumulated globally as evidence that herbivores historically have been limited by predation, and they provided cases of the direct effect of predator removal on herbivore populations, such as the mule deer (*Odocoileus hemionus*) herd irruption on the Kaibab Plateau.[7] This deer population, which lived on the northern rim of the Grand Canyon in Arizona, increased sharply almost immediately after extirpation of wolves (*Canis lupus*) and substantial removal of cougars (*Puma concolor*) in the 1920s, with estimates of deer numbers ranging from 4,000 in 1908 to 30,000 from 1923 through 1930. The powerful ramifications of HSS' hypothesis—that predators influence community dynamics at all trophic levels—inspired vital new hypotheses about community ecology.

Trophic cascades are based on HSS' three trophic levels (an odd number), in which a top predator consumes herbivores at the next lower level, and that in turn affects vegetation at the next level below. These cascades are essentially the indirect effects of predation. Direct effects occur via a predator killing prey, while indirect effects are mediated by a third species. An example would be the indirect effects of sea stars on vegetation in the rocky intertidal zone caused by changes in mussel density via predation. In some systems indirect effects of predation can also arise as a result of behavioral changes by prey in response to the threat of predation, which we will further explore in this chapter. In all cases, indirect and direct effects of predation interact to structure ecosystems.

Aldo Leopold was one of many who contributed to this more enlightened perspective, as was Elton. Although Elton and Leopold identified trophic cascades in primarily terrestrial systems, such as the Kaibab Plateau and northern Mexico, today the majority have been observed in aquatic systems. The sea otter (*Enhydra lutris*), sea urchin (*Strongylocentrotus polyacanthus*), and kelp (*Laminaria* spp. and *Agarum cribrosum*) cascade reported by marine ecologist James Estes in Alaska provides a classic example. When Vitus Bering explored the North Pacific in 1741 he found shorelines teeming with sea otters. By 1911 they were nearly extinct. Enough remained in sheltered rocky pockets along the coast that by the 1960s the otter population had recovered in places. Otters have a varied carnivorous diet and prey heavily on sea urchins. In areas without otters Estes found herbivorous sea urchins thickly carpeting the ocean floor—and no kelp. On reefs with otters he found a lush, green kelp forest and low numbers of sea urchins.[8]

Top-Down versus Bottom-Up

Not everyone bought the green world hypothesis. William Murdoch's counterargument, termed the plant self-defense hypothesis by conservation biologist John Terborgh, suggests that food (bottom-up control) has the strongest influence, that the world may be green because not all plants are palatable to herbivores, and that predators are unnecessary for ecosystem regulation. Or, as evolutionary ecologist Stevan Arnold puts it, the world is green, but that doesn't mean it's edible.[9] Murdoch asserted that food shortage and plant defense strategies may be regulating herbivore numbers. He hypothesized that while plants are essential for the survival of the trophic levels above, the reverse is not true.[10] He and other critics of the green world hypothesis, such as ecologists Donald Strong and Gary Polis, suggested that HSS' failure to address the full complexity of systems, which includes omnivory, weakened their hypothesis. Some have identified HSS' tri-level trophic model as a hypothetical construct because in the real world food webs are not so tidy and can have fewer or more than three levels.[11] Still others, such as conservation biologist Michael Soulé, believe that top-down versus bottom-up, like all dualisms, is false, because the natural world is complex and bottom-up forces (nutrient flow) interact with top-down forces (the effects of predation). Other scientists, such as Rolf Peterson, concur; in the wolf–moose–balsam fir system he studies in Isle Royale National Park, it's never one or the other but a synergy of the two.[12]

HSS' hypothesis was based on a terrestrial environment mostly involving insects and their predators, with the Kaibab mule deer irruption the sole example presented of a mammalian system. This provoked a strong rebuttal by wildlife ecologist Graeme Caughley, who suggested that because the factors that may have resulted in this irruption were "hopelessly confounded," a case study of the Kaibab provided an ineffective example of top-down control.[13] These factors included herbivory by livestock and unreliable deer counts. HSS based evidence for top-down effects on things that control insects and on the connections between insects and plants. They observed that plants flourished in the absence of insects when insects were kept down by predation. Researchers working in various marine and mammalian systems subsequently tested the green world hypothesis. In all systems, whether pertaining to insects or mammals, they found a strong positive correlation between predator removal, plant community simplification, and reduced energy flow, which goes back to Paine's dictum about being able to change the world by removing one species.

Ecologists Lauri Oksanen and Stephen Fretwell proposed that if we treat trophic levels as units, systems having four or more trophic levels may have more than one level representing predation. In this scenario a predator at the fourth level will dominate a predator at the third level, and this will release the herbivore population at the second level from predation, causing its numbers to increase. This in turn will cause overgrazing of plant communities at the first level. Trophic cascades in which a top predator controls its herbivore prey via top-down forces will therefore always have an odd number of trophic levels. Removal of the top trophic level in such systems will have a radical effect on lower levels, causing herbivore irruption and overconsumption of vegetation. In any food chain, energy flow alternates; in odd-linked systems, plants will be limited by resources available to them (top-down control); in even-linked systems, plants will be limited by grazers (bottom-up control). Thus systems with an odd number of levels will be green, while systems with an even number of levels will be brown or barren.[14]

In the 1990s Estes saw a three-level trophic system flip into a four-level system in coastal Alaska. The apex carnivore *Orcinus orca*, the killer whale, began preying on sea otters, reducing their numbers significantly. Killer whales added a fourth level, reversing the flow of energy from top-down to bottom-up. The resulting cascade rippled through the North Pacific, causing an upsurge in sea urchin numbers and kelp consumption and denuding the ocean floor.

Researchers have studied how the green world hypothesis holds up under different levels of ecosystem productivity, called net primary production. In 1981 Oksanen and his colleagues developed the exploitation ecosystems hypothesis (EEH). In their model herbivory is greatest in relatively unproductive environments, with predation more important and the impact of herbivory reduced as ecosystem productivity increases. Deserts and arctic regions provide examples of systems where plant production is so scant that it can fail to support herbivores. At slightly higher productivity levels, such as on prairies and savannas, an ecosystem becomes able to support a consumer trophic level, with herbivory increasing as productivity increases. As productivity continues to increase, as in boreal forests, plant *biomass* (the total mass of living matter of a particular type) increases, herbivore biomass increases, and ecosystems become capable of sustaining a third trophic level—the predators—with this level controlling herbivores. Because it uses environmental productivity as a key driver of ecosystem dynamics, EEH incorporates both bottom-up and top-down forces.[15]

The Keystone Species Concept

The keystone species concept lies at the heart of the HSS debate. When Robert Paine introduced it in 1969, he envisioned its mechanisms as a dominant predator consuming and controlling the abundance of a particular prey species and a prey species competing with other species in its trophic class and excluding them from the community. As long as one keeps these two processes in mind, the whole idea of keystone species fits into place—like the keystone of an arch. And when the keystone is removed, arches and ecosystems fall apart. This mechanism explains the ecological collapse Paine observed on Mukkaw Bay. One year after he began removing *Pisaster*, seven of the fifteen species he had inventoried at the start of his star-throwing experiment were gone, with the others declining rapidly.[16] However, some scientists suggest that ecosystems may not exactly "collapse" when a keystone predator is removed. An ecosystem thus altered continues to function, albeit in a different manner, by moving into what is referred to in ecology as an *alternative stable state*. We will explore ecosystem shifts caused by removal of keystone predators further in chapter 3.

The keystone concept has been applied to a variety of species, from large carnivores to herbivores and even plants. And it has created disagreement among ecologists, who have questioned this term's usefulness and generality. In response to mounting criticism of the overly broad use of this popular term, marine ecologist Bruce Menge defined a keystone species as "one of several predators in a community that alone determines most patterns of prey community structure, including distribution, abundance, composition, size, and diversity." Keystones selectively prey on dominant prey, have large body size relative to prey size, and are highly mobile, with a large foraging range. More recently Michael Soulé and his colleagues proposed the term *strongly interacting species* for those that have a strong ecological effect on communities.[17] Next we will see how fear drives some of these effects.

The Ecology of Fear

Mid-May in Glacier National Park, Montana, is not a time or place for the fainthearted. The gunmetal sky was beginning to spit snow at me and my field technician, Dave Moskowitz, as we hurried to finish a track transect before a late-season blizzard broke out. All day we had been hearing wolves howling. It had been difficult to pinpoint their location over the rising wind, but it seemed as if they were moving along a benchland one-tenth of a mile east of us. Hearing them like this provided a powerful reminder of the wildness of the system I was studying.

My work involved putting in fifty-seven miles of track transects in Glacier National Park's North Fork, arguably one of the most intact systems in the lower forty-eight states. This place harbored a full suite of large predators in one of the highest densities found south of Alaska, as well as abundant elk (*Cervus elaphus*) and deer (*Odocoileus* spp.). An expert tracker, Dave was writing a field guide to tracks of the Pacific Northwest. I was fortunate to have his help. Using track transects in which we were measuring all occurrences of elk, deer, moose (*Alces alces*), wolves, cougars, coyotes (*Canis latrans*), and bears (*Ursus* spp.), I was mapping their interactions and determining game densities in a system with so many predators. I was also measuring predation risk, to see whether wolves and other predators influence elk movements by causing them to avoid areas with escape impediments, such as downed wood and thick brush. In a 1999 article wildlife ecologist Joel Brown noted that the nonlethal effects of predators can be ecologically more important than the direct mortality they inflict.[18] Evidence I'd found thus far created a compelling picture of the same sort of ecologically complete system Aldo Leopold had observed in 1936 in Mexico. I was also seeing patterns sharply etched on the landscape. These elk were spending more time on flat, open ground and less in riskier terrain, where they might have had more difficulty escaping predators.

As we raced to finish before the weather deteriorated further, we encountered a bachelor herd of a dozen skittish elk at a distance of fifty yards. They had shed their antlers and sported velvety antler buds. I gauged their ages from their body size and behavior. The younger bulls nervously ran around in circles. The mature bulls eyed us warily but stood their ground. Eventually they left, the young bulls taking cues from their elders, moving in the elegant head-high trot characteristic of their kind. As they vanished into the inky conifers below the benchland, I was left with an ominous feeling in the pit of my stomach. Something about their behavior made me uneasy, but I couldn't put words to it.

We finished just as the blizzard hit. When I looked back toward where we began the transect, I encountered a sight so unexpected, so shocking, that initially I thought I was imagining it. For there, conspicuous even through a scrim of falling snow, lay a fresh bull elk carcass. The animal looked impossibly huge. The wind carried clouds of steam from its gaping belly. I was amazed that we hadn't heard the takedown, even over the moaning wind. One minute the dead elk wasn't there and the next minute it was, having met death in an area I had moments earlier characterized as having very high predation risk.

I investigated the carcass—an old bull surrounded by crimson wolf tracks on snow, its flesh still warm beneath my hands. Already the wolves had removed most of its hindquarter meat and some organs. While Dave photographed it I looked up, my senses sharpened by the coppery scent of blood and this primal encounter with the ecology of fear. A fresh carcass soon draws cougars and bears; I lingered only sufficiently to record location coordinates.

The next morning a young grizzly fed on the carcass. I watched it patiently for an hour as it removed much of the remaining meat, made a vain attempt to haul the still heavy carcass up the bench, and then lumbered away to sleep off its meal. All the while a Steller's jay perched on a downed log just to the side of the carcass. It repeatedly tried to scavenge meat, each attempt met by a growl and swat from the grizzly. Later that day I observed a coyote trotting through the area, balancing a purloined elk leg in its mouth. Bears emerge from their dens in April, ravenously hungry. One of their survival strategies involves searching out wolf kills. In this park, as in Yellowstone National Park, many wolf kills end up usurped by grizzlies. Carcasses such as this one show how apex carnivore predation can support a wealth of species, from grizzlies to jays to coyotes.

Six months later, in mid-November, I worked in Waterton Lakes National Park. The wind blasted across the Alberta prairie, nearly knocking over my tripod and spotting scope. I steadied them with gloved hands and tucked my chin deeper into the collar of my down parka. It felt more like mid-January in this extreme landscape, where most years the only month I didn't experience snow at my field sites was July. I was out there using yet another method to determine whether elk fear wolves. Conservation biologist Joel Berger, who has done global research on the fear of predation, believes this phenomenon underscores more fundamental questions—the meaning of fear itself and how it can affect ecosystems.[19]

My teenage daughter Bianca had joined me in the field. We were watching a herd of approximately four hundred elk cows, which stood on a high benchland on the southeastern border of the park. Most had their heads up, scanning the landscape rather than eating. They skittishly grazed on tawny dried remnants of prairie grasses that poked up through the thin snow, taking quick bites, looking up for some long moments before stealing another mouthful of food.

"What's up with the elk?" asked Bianca.

"You'd think there were wolves nearby," I said.

"You think?" she asked.

Soon the wolves would oblige us with an answer.

The ecology of fear has deep roots. Staying alive during the early Pleistocene epoch involved escaping large creatures with sharp teeth and claws. This meant that prey species evolved behavior driven by survival. Vigilance—time spent head up, looking for threats—is essential for survival in systems with top predators, but it comes at the expense of time spent eating. For the past two years I had been doing focal animal observations on elk. This involved watching one animal at a time, recording how long it spent with its head down feeding versus head up, scanning for predators. I had categorized my study sites as areas of high, medium, and low wolf presence. I wanted to know whether fear varied on the basis of wolf density and, from that, to learn how many wolves would be enough to trigger changes in herbivory patterns—a trophic cascade. Would one pack passing through an area occasionally, but not denning there, have the same effect as two very large packs that had produced multiple litters of pups in one year? How about one pack that kept hanging on despite losing half of its members annually as a result of human-caused mortality in an area where it was legal to shoot wolves outside the park? The elk I observed on that blustery November afternoon in my medium wolf density area were far more skittish than usual.

All at once five black shapes crested the bench, fluidly trotting through the elk. Wolves. We'd seen their tracks earlier that day, pressed into a skiff of snow on the Chief Mountain Highway. Few humans visited the park between late fall and spring; predators and other wildlife adapted to this by increasing their use of park roads that are closed to vehicle traffic in the off-season. Even through the blowing snow I could see that these wolves were muscular and well fed. They moved comfortably through the herd, shifting into a slow lope, tails high. The elk parted as the wolves passed, and then they regrouped a short distance away, heads up, bunched more tightly for safety. The wolves didn't stop but continued on their way, disappearing over an eskerine ridge. It was a long while before the elk settled and resumed feeding, and their vigilance level remained high for hours. Their earlier restlessness made perfect sense.

Aldo Leopold was among the first to observe the behavioral effects of lack of predation on his own land. In 1935 he bought an abandoned farm in southwestern Wisconsin, to use as a hunting reserve. He subsequently dubbed the farm "the shack." This land, now known as the Leopold Memorial Reserve, lies forty-five miles north of Madison, on the southern edge of Wisconsin's sand counties. The Leopold family spent every weekend there, restoring the land. Between 1939 and 1940, in his shack journals Leopold noted the effects of deer herbivory on herbaceous plants (plants whose leaves and stems die down to soil level at the end of the growing season) and trees he had planted on his land, which included oaks (*Quercus* spp.) and aspens (*Populus tremuloides*). He commented that some species were being nipped down to eighteen inches in height. In a game survey he also documented how humans had by then eliminated wolves from much of Wisconsin. Deer had exploded in northern Wisconsin, from several hundred in 1920 to at least 100,000, causing game managers to formally acknowledge the problem. Although things were not this bad at the shack, Leopold noted ongoing plant damage caused by deer, which in the absence of wolves calmly stood their ground and browsed young saplings down to nothing.[20]

There is a saying that the more things change, the more they stay the same. And an older Romanian proverb from the Karpaten states that where wolves go around, the forest grows. Both pieces of folk wisdom came to mind when I visited the shack to see whether the current aspen growth pattern and deer behavior would fit Leopold's historical observations. It was mid-April as I drove along the rural road that runs through the Leopold Memorial Reserve, noting small herds of deer standing around with their heads down, eating shrubs. I spent the next few days examining the amount of browsing on aspen sprouts.

One day Aldo Leopold's daughter Nina Leopold Bradley joined me in the field. Her chocolate Labrador retriever, Maggie, ran glad circles around us as we examined the aspens around the shack for evidence of deer herbivory. Almost all the aspens below browse height (the height a deer can reach to eat) featured chisel-pointed ends where deer had bitten off the apical stem, the dominant growth bud. Many had zigzagging trunks, where they had been browsed and had healed, and then had grown in a different direction. I held my measuring rod to an aspen less than three feet tall and counted its browse wounds—eight in all, each marked by a crook in its trunk. I showed Nina how telltale signs on the aspen allowed us to distinguish browsing from disease, because the latter made the trees' growth tips atrophy. All aspens can sustain moderate browsing, but these bonsai aspens looked stunted and shrublike. With chronic herbivory they would eventually die. Indeed, we found many that had succumbed in this manner.

Nina and I reflected on how little things had changed at the shack since her dad's era. As we continued to walk she recalled his observations about the effect of wolf removal on deer behavior, and how deeply this awareness affected him. "My father always said it all had to do with relationships. But he couldn't convince managers of that. He was even unable to convince some of his best friends. He had found stacks of dead deer as big as a house in northern Wisconsin, and his colleagues would not vote for a doe-hunting season. I think we have the same problem today. I do not think that people, even at the highest level, quite understand the interdependence of all of these issues. We still have too many deer, we still have hunters thinking we don't have enough deer, and we still have no wolves here."[21]

So how do these relationships work? Let's say you are a white-tailed deer foraging on the Leopold reserve. There are no large predators in the forest that could threaten you. So you feed steadily on shrubs and grasses, looking up only to interact with others of your kind or search for food. Now let's say you are a white-tailed deer foraging in the North Fork. You take a bite and look up, sacrificing food for safety, highly vigilant. You are living in a landscape of fear, where your ability to survive depends on your ability to detect and escape predators as well as obtain food. The resulting stealth and fear dynamics—and the relationship between top predators and their prey—have profound ecological implications.

Risky Business: Predation and Resource Selection

Predator-prey interactions have two components: predators killing prey and predators scaring prey. While the lethal effects of predation are well documented, nonlethal effects may have equally strong consequences. Joel Berger tested this by tossing snowballs imbued with predator scents, such as wolf urine and grizzly bear feces, at ungulates. In addition to pungent snowballs, he experimented with tape recordings of predator sounds (lion roars and wolf howls) and neutral sounds (water and monkeys). Where wolves had been absent for decades, such as in Rocky Mountain National Park, elk responded to the snowballs or predator sounds with some curiosity, but none became alarmed or ran. In Denali National Park and Preserve, where wolves had been present for many years, ungulates responded by becoming hypervigilant. Berger and colleagues continued this work on a circumpolar scale, working in Greenland and Siberia, where predators had long been present, and finding similar results with moose and caribou (*Rangifer tarandus*). Beyond individual responses, Berger wanted to know how prey animals acquire knowledge, how fear is transmitted, and how current behavior can help unravel the ambiguity of past extinctions and contribute to future conservation. Ultimately his work will help shed light on how predators shape prey behavior and landscapes.[22]

Research about predation risk has the potential to inform human choices about which landscapes can be allowed to harbor dangerous animals. Berger and colleagues found that in Wyoming moose increased vigilance behavior in the presence of grizzly bears, keeping their heads up longer and staying on the move to avoid predation. This reduced browsing on willows, enabling the willows to flourish, thus improving habitat for songbirds and increasing biodiversity. Awareness of these landscape-scale effects can be used to make management decisions about grizzly bears, perhaps allowing them to expand their ranges.[23]

Predation is the main driver of fear in prey because it can lead to death. Fear of predation involves a response to predation risk, whereby prey react to predator presence—or even to the mere threat of it. Fear causes the adrenal glands to secrete adrenaline, a short-acting substance that prepares the muscles and brain for flight. It also produces cortisol as part of an animal's long-term response to chronic stress. An elk uses all its senses to evaluate the threat of predation. Its ability to assess and control its risk of being preyed upon strongly influences habitat selection and feeding decisions.[24] Prey animals establish an optimal baseline level of vigilance in the absence of direct evidence of predator presence. Individuals who successfully balance the benefits of risk avoidance against energy costs (missed opportunities to eat) have a greater chance of survival.

This response is not limited to large mammals. Working with animals at the opposite end of the size spectrum, Oswald Schmitz found a behavioral trophic cascade consisting of spiders and grasshoppers. The top predator he studied, the nursery web spider *Pisaurina mira*, preferentially preys on the grasshopper *Melanoplus femurrubrum*. In the absence of spiders, grasshoppers selected a diet composed almost entirely of grass rather than forbs (flowering plants that are not grasses). In this famed experiment Schmitz glued spiders' mouths shut to render them unable to prey on grasshoppers. In the presence of spiders with glued mouths, grasshoppers nevertheless reduced their feeding time and preferentially ate forbs, which provide greater cover and safety from predation. This shift resulted in a trophic cascade.[25]

Behavioral adaptations are complex and variable and show an evolutionary relationship to landscapes. In large mammals, behavior that evolved over thousands of years underlies trophic cascades mechanisms. Elk originated in Asia, on high grassland steppes. They colonized North America about 10,000 years ago, crossing the Bering land bridge. Lacking competition from other elk species in North America, they spread widely across many habitat types, from Pacific Northwest rain forests to sagebrush deserts.[26] Long-legged cursors, elk run with their heads up and a straight-legged gait (as opposed to bounding). They escape predators via rapid and sustained flight, an adaptation found in ungulates from open plains with low flight impediments. On landscapes with both open and closed habitat structure, they may use a combined strategy of hiding in forest cover to lower predator encounter rates and seeking open terrain, such as grasslands, where predation risk may be reduced.[27]

Recent studies have examined factors that can render prey more vulnerable, such as differences in ungulate grouping behavior and terrain features. In Banff

National Park, landscape ecologist Mark Hebblewhite found predation risk lowest in small groups of elk, with groups larger than twenty-five having the highest probability of being preyed upon by wolves, possibly because they are more likely to contain weak or sick individuals and are easier for predators to detect.[28] In Yellowstone, ecologist William Ripple developed his predation risk hypothesis while sitting on a high terrace in the Lamar Valley, where he noticed patchy willow growth. Out in the open, willows were browsed down, but where there were visual or terrain obstacles, willows flourished. He and a student, Joshua Halofsky, proceeded to measure elk behavior and found heightened vigilance in areas with escape impediments. Elk spent more time with their heads up, scanning for predators, in these areas, behaving more skittishly than when they were assured of a clear escape route.[29]

According to Douglas Smith, leader of the Yellowstone Gray Wolf Restoration Project, the concept of predation risk eludes easy definition. For example, an area where wolves take down prey after a long chase may not necessarily be the site of greatest predation risk. That may actually be the site where prey first encounter predators. Additionally, Smith notes that most wolf kills occur between dusk and dawn. Because elk and other ungulates have poor vision, obstacles to their viewshed may not play a significant role in the dynamics of predation risk.[30] Presence of other predator species complicates matters. Wildlife biologist Kyran Kunkel found that avoidance of one species, such as the cougar, which hunts by stealth, makes prey more vulnerable to another, such as the wolf, which runs down its prey.[31]

Elk have a sophisticated response to predation risk that includes gathering in larger groups in open areas. Landscape ecologist Matthew Kauffman and colleagues found that open areas enable wolves to detect prey more easily and thus present greater predation risk. Wildlife ecologists Stewart Liley and Scott Creel found that elk adjust their vigilance in response to the size of their group and the type of immediate threat they face from wolf presence, with environmental variables such as obstacles having a secondary influence on vigilance.[32] Some researchers recommend that trophic cascades studies incorporate radio-collar data to measure behavioral predation risk (i.e., wolf presence). According to Smith, the complexity of these interactions merits deeper investigation.

I ended up putting in 150 miles of track transects in the Crown of the Continent Ecosystem and doing 700 focal animal observations. In doing this work I found compelling evidence of a trophic cascade. Where wolf density was high, elk avoided areas with debris and other escape impediments. Most carcasses and the greatest amount of wolf sign, such as tracks and scat, occurred in thick forests, debris, ravines, and riverbanks, which I had characterized as high predation risk sites. My focal animal observations suggested that the more wolves there are in a landscape, the more wary elk become. This response may be triggering cascading effects in this ecosystem, enabling aspens to grow above browse height. Indeed, the ecology of fear may be behind the changes at my home, where shrubs and trees have reclaimed the meadow after wolves returned and deer, to stay alive, have had to act more like deer and less like livestock. These pervasive effects influence even small, nonkeystone predators, as we shall see.

Mesopredator Release

It was late May in Waterton Lakes National Park and the snow had just melted, the matted grasses still a winter-killed brown. My field crew and I were putting transects into an expansive rolling grassland dotted with aspen stands. My field technician Blake Lowrey was on point that day, doing dead reckoning with a compass and pulling the transect tape due east through a copse of stunted aspens, making good progress. All at once he stopped and said, "What's that smell?" I had cautioned everyone to be careful around carcasses because of bears. At the head of the transect line I found a dead coyote on a well-used game trail. This relatively fresh coyote carcass had been there for maybe one or two days. It lay on its back, limbs outspread and neck outstretched. Its throat had been ripped out and it had been eviscerated. No other flesh had been removed. All around it lay evidence of the perpetrator of this carnage: wolf feces and tracks. The coyote appeared to have been a young adult male in relatively good health that had perished because it had had the misfortune to come upon a wolf. In most systems wolves make it their business to kill coyotes. This particular carcass had been left on a primary game trail as a grisly marker and warning to other coyotes that wolves rule this system—they are the apex predator.

Two weeks later, in Glacier National Park, we found another coyote in the same position, also on a game trail, its throat and guts ripped out, no other flesh missing, wolf scats and tracks all around. By now my crew had become sufficiently accustomed to carcasses to find this fascinating. We took a break and discussed the possible pattern here—the powerful signature wolves were leaving on this landscape.

Wolf-coyote enmity is not new. Wolves recolonized Isle Royale National Park in the early 1950s. Within two years they had wiped out the resident coyote population.[33] When wolves were functionally extirpated from Yellowstone, wildlife biologist Adolph Murie noted a corresponding steep rise in coyote numbers, which began to form larger packs and hunt deer. Smith considers what happened next to Yellowstone's coyotes one of the best stories to emerge from the mid-1990s wolf reintroduction. After the wolf's return, coyote numbers dropped by as much as 50 percent overall and by 90 percent in core wolf pack territories. To survive, they formed smaller groups and spent more time in the interstices between wolf territories and nearby roads. There Smith found coyotes killed by wolves in a similar manner as I'd observed. Most of the pre-wolf coyote population had occurred in packs. Since wolves, half the coyote population has consisted of what Smith calls "floaters," unaffiliated coyotes with higher survival rates. Breeding coyotes have the highest mortality because they are easiest for wolves to find and kill, since their behavior is more predictable and they live in territories.[34]

One of the most powerful indirect effects of predation involves mesopredator release. Defined as medium-sized predators, mesopredators are controlled by top predators—often by direct mortality, as we have seen, but also via competition for shared resources. Humans commonly remove keystone species to protect economically valuable big game from predation. For example, upon removal of the wolf from the Endangered Species List in the northern Rockies, the state of Idaho proposed to eliminate 40 percent or more of its wolves to help create more elk for humans to hunt. This type of action causes mesopredators, such as coyotes, to increase and puts abnormal pressure on smaller species, such as game birds, which decline and can become extinct.

In the mid-1980s David Wilcove investigated the effects of human land use on songbirds. He studied small woodlots in rural and suburban sites in

Maryland and larger forest patches in Tennessee in Great Smoky Mountains National Park. The biggest tract of virgin forest in the eastern United States, this park retained forest-dwelling mammals and birds long extinct in central Maryland. Wilcove wanted to test the effect of mesopredator release on songbird nest predation—what can be thought of as the empty nest hypothesis. To do this he filled experimental wicker nests with Japanese quail (*Coturnix japonica*) eggs and placed them in forest locations ranging from the midcanopy to the understory, to reflect native birds' nesting habits. One week later he measured the percentage of experimental nests raided by mesopredators. He found higher rates of nest predation in small woodlots near human communities because these areas had higher populations of raccoons (*Procyon lotor*) and squirrels (*Sciurus* spp.) and few, if any, large predators, such as cougars and bobcats (*Lynx rufus*). This led Wilcove to link nest predation to mesopredator release.[35]

As Wilcove was finishing his work, on the opposite side of the country Michael Soulé began to study how coyote decline affected the ecology of Southern California's chaparral canyons. He lived north of San Diego, a few miles from the coast, in an area about to be developed, which would fragment wildlife habitat. Soulé spoke with the developers and suggested preserving some open lands and maintaining connectivity corridors between them. The developers wouldn't listen, so he got mad and began a research project in the chaparral, also called coastal sage scrub habitat.

Soulé and his students investigated what happens to birds in isolated patches of habitat. This landscape-scale project involved surveying thirty sites for the presence of seven species of chaparral-dependent birds, beginning in a two-acre habitat patch created forty years earlier by development. He recalls sitting in that patch for an hour, listening and looking for birds, and finding it silent except for mockingbirds (*Mimus polyglottos*) and jays (*Aphelocoma insularis*)—generalist species that do not need chaparral. As he noted the absence of quail (*Coturnix* sp.), roadrunners (*Geococcyx californianus*), sage thrashers (*Oreoscoptes montanus*), and other chaparral-dependent species, he realized that there had been significant extinction in some of these old patches.

Among the variables Soulé used to analyze his data were coyote presence or absence. Coyotes act as keystone predators in some systems and mesopredators in others. In Yellowstone, where wolves are keystones, coyotes take the mesopredator role. In Southern California, where there are no wolves and probably never were any, coyotes take the keystone role. Soulé hypothesized that falling coyote numbers in an area being developed by humans would result in the release of native and exotic mesopredators such as raccoons and housecats.[36] His analysis showed that patches with coyotes contained more chaparral-requiring birds and, on the other hand, patches without coyotes had fewer such birds. But his epiphany came when he realized that this was actually a cat effect. He owned cats and knew they had trouble surviving around coyotes because of predation. So what he was observing with the birds was the indirect effect of coyote absence (more cats) and thus a mesopredator release.[37]

Ten years later Soulé's student Kevin Crooks deepened this study by radio-collaring cats. His data very graphically demonstrated the relationship between lack of coyotes, increased cat movements, and reduced populations of chaparral-dependent birds. Soulé's mesopredator study had been based on a statistical correlation showing the trophic cascade coyotes were producing, but Crooks confirmed this empirically using radio-collars. Accordingly, Crooks and Soulé concluded that coyotes were ecologically beneficial because they controlled mesopredators that preyed on birds while rarely preying on birds themselves.[38]

Since this study, mesopredator release has been identified in the Dakotas, where coyote absence caused the red fox (*Vulpes vulpes*) population to surge, making survival far more challenging for prairie ducks. Similarly in Texas, coyote removal led to an increase in five species of mesopredators and a decrease in game birds. In these cases mesopredator release reduced biodiversity and demonstrated the ecological importance of the alternative food web pathways created by keystones. These relationships raised scientific awareness of what may be at stake ecologically when we lose a keystone.

Remembrance of Things Past: Megafaunal Extinctions

We live in a world of losses. Cave paintings dating back to the upper Paleolithic period, between 30,000 and 10,000 years ago, depict creatures long gone from the northern latitudes—mammoths, horses, cave bears, rhinoceroses (Rhinocerotidae family), and saber-toothed tigers.[39] Today we are drawn to these haunting images of a lost world. What exists today in terms of large fauna is but half of what once was.

A *guild* is a group of species that have the same role and coexist in an ecosystem. For example, wolves, cougars, and grizzly bears form the predator guild. Predator guild extinctions have far-reaching associations that go back beyond near time into the distant past. These effects invite us to take a closer look at a lost world shrouded in mystery.

Most large mammals, termed *megafauna* (animals weighing more than 100 pounds), became extinct in the past 50,000 years, during the late Pleistocene epoch, when *Homo sapiens* colonized the earth. We have lost more than 150 genera of megafauna in this time span. Paleontologist Paul Martin and others believe that spear-wielding humans brought this time of mammalian giants to an end.[40] But some scientists, such as R. Dale Guthrie and Aldo Leopold's youngest daughter, Estella Leopold, a renowned palynologist, believe that climate change and disturbance also may have had a major influence on these extinctions. To illustrate the role of climate on extinction, Estella recounted her experiences while doing research in the Mojave Desert:

> We were in Death Valley, in the third year of our study, measuring the impacts of drought. It was a barren landscape, with little growing and almost no wildlife. And then it started to rain during the fourth year, and everything recovered. Creatures we thought were gone returned, like quail, tortoises, snakes, and lizards. And then there was another drought, and the plants and animals disappeared again. I had been studying the paleosediments of these valleys, using drill cores, doing pollen analysis, and finding out that these kind of jerks went on down through time two million years. The bottom of the core, which is bedrock, is a long section of nothing, and then you begin to see the plants of the Sierra Nevada showing up, and then you begin to see plants disappearing and reappearing. The picture you get from these cores is the same as you get if you sit on the edge of the lake today, after one of these droughts, the intermittent wobbles. You see birds coming and going, flowers com-

ing and going. This has been happening for nearly two million years. It very dramatically tells you about ecosystems and their evolution.[41]

While understanding causes and effects of megafaunal extinctions will require a deeper knowledge of climate change and prehistoric human land use chronologies, everyone agrees that they had major impacts on plant communities. Imagine a world with five times as many herbivores as we have today and twice as many large predators. In North America extinct megafauna include a giant armadillo, three species of giant ground sloth, the Columbian mammoth, a mastodon, a giant tortoise, a camel, a saber-toothed cat, an American lion, and a woodland musk ox.[42] With the exception of Africa, megafaunal extinctions occurred globally, leaving ecologists to wonder what it means when prairies grow silent in the absence of roaring lions and thundering bison hooves.

Although we will never know the impact this had on Pleistocene vegetation, we know from the few places where echoes of such assemblages still exist that these effects can be tremendous. In Africa, elephants (*Loxodonta africana*) mainly browse woody species, changing the landscape by pushing over, breaking, or uprooting trees, creating openings in the forest and helping maintain grasslands. Similarly, *Hippopotamus amphibius*, which primarily feeds on grasses, transforms tall grasslands into a mosaic of short and tall patches that support a rich variety of species. Pollen and fossil records suggest that grazing megaherbivores, in the form of mammoths and ground sloths, together with browsers such as mastodons, may have helped maintain the open parkland vegetation that covered much of North America. Conservation biologists have been profoundly struck by the implications of Paul Martin's work. Loss of large-bodied terrestrial and aquatic fauna may have had a huge impact on systems and suggests many trophic cascades that formerly arose from top predators have vanished.[43]

The ultimate keystone predator, humans alter their environments by eliminating species and modifying ecosystem structure and function, thereby contributing to extinction, altering evolution itself. Megafaunal extinctions have had enduring effects, simplifying ecosystems and eliminating large predators such as the dire wolf and saber-toothed tiger, large herbivores such as elephants and giant sloths, and the suite of large scavengers supported by these predator-prey interactions.[44] Ecosystems have been truncated or decapitated by the loss of larger animals. Beyond evolutionary entanglements, when one views these extinctions through trophic cascades glasses, the profound ecological wreckage humans have inadvertently wrought on this planet begins to become apparent.

What does megafaunal extinction mean in terms of the distribution and characteristics of existing plant communities? It's about pattern and process. All the green growing things on this earth evolved over millennia, adapting in response to changing environmental conditions. Systems became simpler toward the end of the Pleistocene because of the loss of megaherbivores. The remaining top predators gained a more powerful keystone role, and plant distribution again adapted to new herbivory patterns.

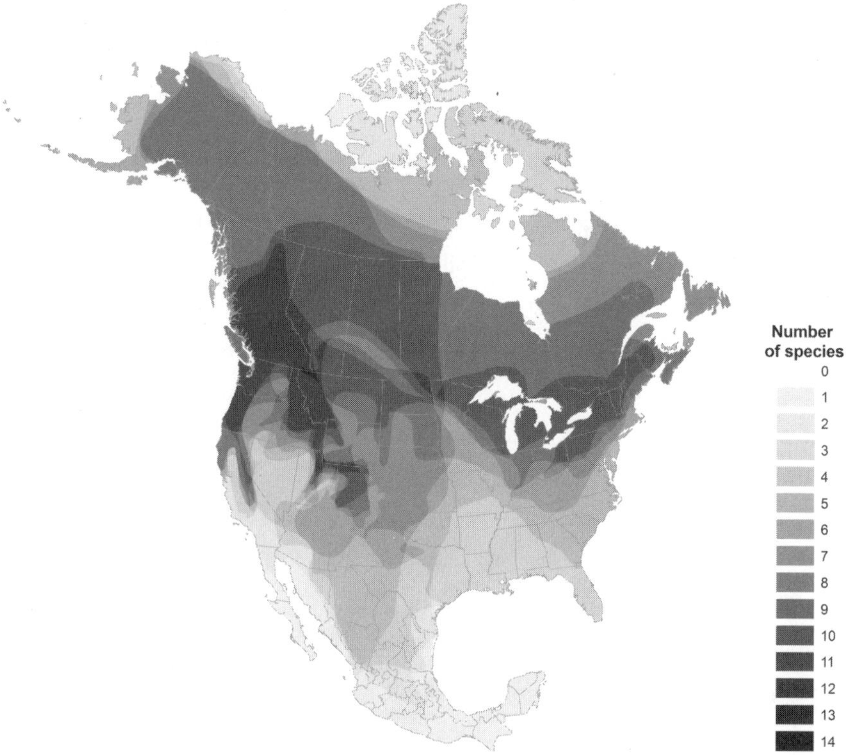

Figure 2.1. North American Historical Carnivore Ranges

Species loss has continued into the present day, driven by a variety of factors, including climate change and overharvest by humans. Andrea Laliberte and William Ripple found that some North American carnivores have experienced range contractions to less than 20 percent of their early eighteenth-century range.[45] This has resulted in irruptions of ungulates unchecked by predation, followed in many systems by reduced diversity of plants and wildlife such as songbirds.

Landscapes evolve in time and space. These mechanisms leave patterns in a landscape. The elk I observed on that eskerine benchland in Waterton, heads high, taking cautious, quick bites of grasses and aspens and then looking up to

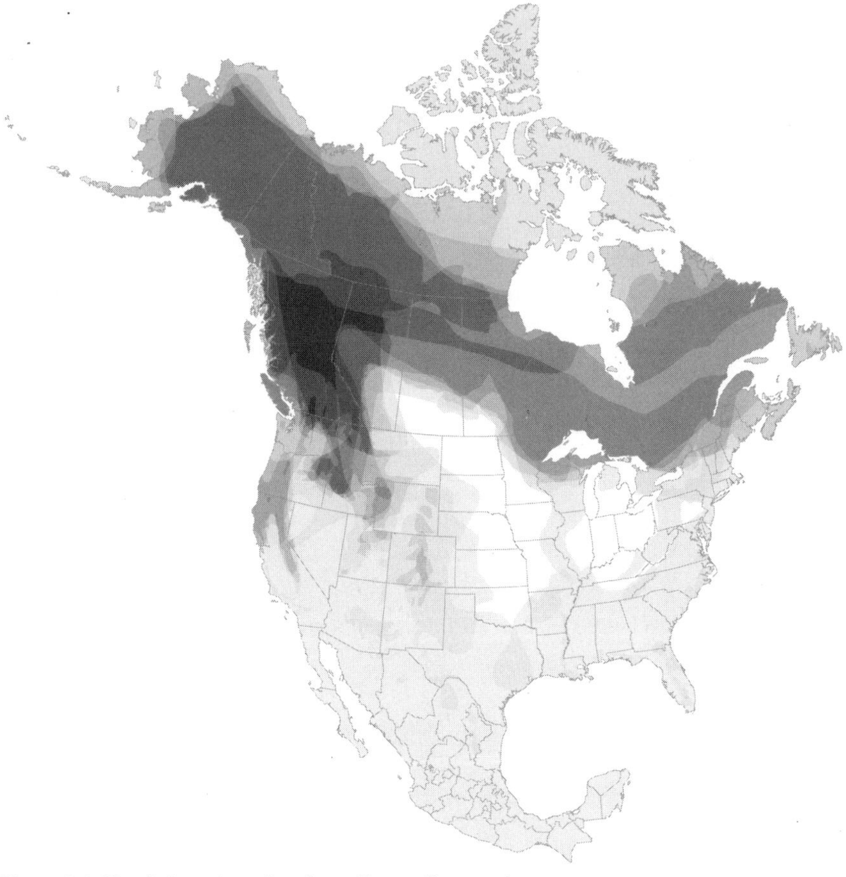

Figure 2.2. North American Carnivore Range Contractions

scan for wolves, are shaping the way things grow, creating patterns with their behavior. Landscapes tell stories, if we listen. The trophic cascades mechanisms that underlie this behavior can teach us much of what we need to know in order to conserve biodiversity and create more resilient terrestrial and aquatic ecosystems.

CHAPTER THREE

Origins: Aquatic Cascades

I stood on Yaquina Head on Oregon's central coast, leaning into the cold, salty wind, bundled in GoreTex and polar fleece. The spring storm building in the west painted ocean and sky in shades of brooding gray. A black volcanic rock beach curved south, giving way in the misty distance to a broad, sandy crescent. Cliffs crowned by wind-sculpted firs palisaded the beach to the east. Waves swept the shore at high tide, unleashing their energy on the rocks. Water surged into the more sheltered coves with the ocean's rhythm, ebbing in white rivulets from crevices and channels carved by the surf. I could hear the low rumble of rocks tumbled by waves.

The next afternoon I came out to the shore to find a different world. The sun was out, the ocean a deep azure blue, the wind calm. Harbor seals and shorebirds rested on small islands. Tucked beneath the headland were tide-exposed mussel beds and pools teeming with life. I gingerly stepped among them on basalt cobbles worn smooth by surf, peering in wonder at the marine gardens at my feet. Waves curled and crested beyond the tide pools, kelp fronds and sea grasses undulating in the gentle current. The still, clear tide pool water reflected blue shards of sky and gulls flying low overhead, overlaying these images on the intertidal communities visible beneath the surface. California mussels completely filled many tide pools, their black and russet shells packed into dense beds. Also called mussel barrens because of their lack of biodiversity, these beds were vibrantly hemmed by jade green anemones with purple-tipped tentacles.

The rocky intertidal zone is the narrow band between coast and ocean revealed at low tide twice daily, the result of the sun and moon's gravitational pull on the earth's oceans. Breaking waves and an environment exposed to the air for part of the day shape its ecology. Subject to greater wave action at its lower reach and greater air exposure at its upper reach, this zone features clearly marked patterns. Divided into upper, mid-, and lower regions, it contains distinct communities of organisms, such as mussels, uniquely adapted to various levels of disturbance. The life-forms here depend on constant wave motion to bring them nutrients and oxygen and remove their waste. The splash and spray of waves are part of what creates the diversity of life here, by increasing the amount of rocky habitat available to intertidal species. Just beyond the surf's break lies the nearshore zone, which contains kelp forests and provides a home for seals and otters.

Predation shapes this community no less dramatically than the ocean's power. While many of the intertidal fauna have biologically adapted to the wave action by forming shells and holdfasts, they remain vulnerable to predators. Gulls and sea otters (*Enhydra lutris*) crack their prey's shells open. The ochre sea star (*Pisaster ochraceus*), a slower-moving but no less formidable predator, consumes its prey's soft flesh by extruding its stomach and inserting it into any small opening in the shell. *Pisaster* mostly eats California mussels (*Mytilus californianus*) and occasionally barnacles, snails, limpets, and chitons.[1] Indeed, this keystone predator holds the power to tip this system from mussel-dominated to predator-dominated states.

I wandered the shore observing how this works. Some tide pools contained sea stars, their plump bodies sprawled against one another, ranging in color from bright orange to dusty rose. Iridescent, weblike calcareous laceworks marked their backs, giving them support. Occasionally I found one slumped ominously over a mussel, in the act of consuming its prey. These tenacious predators triggered an escape response in some of their more mobile prey, such as sea snails, which I could see gliding off, an example of the ecology of fear at work in this intertidal world. Tide pools that contained sea stars held far fewer mussels and abundant evidence of predation. Broken shells lay strewn about, their nacreous interiors gleaming. But most notably, predation made room for a panoply of species, fanciful in color, form, and texture: crusty-shelled limpets (*Diodora* spp.); decorator crabs (*Oregonia* spp.), their shells liberally festooned with bits of algae and coral for camouflage; and chevron-striped chitons (*Tonicella* spp.), their oval bodies marked with purple and green ridges. And there were many kinds of marine plants—several types of brown kelp (*Laminaria* spp.), pink coralline algae (*Calliarthron* spp.), yellow-orange sponges (*Mycale* spp.), and scruffy brown "nailbrush" seaweed (*Endocladia muricata*).[2]

The difference between tide pools without sea stars, which held mussel barrens, and those where they abounded, which teemed with species, reasserted itself all along the shore, each a microcosm, each containing a community where trophic relationships were graphically playing out. No less dramatic than the aspen recruitment gap documented by researchers of terrestrial trophic cascades, the rocky intertidal zone held eloquent evidence of a trophic cascade in action and of how a keystone predator, such as *Pisaster*, can create with its presence a significant increase in biodiversity, thereby changing the appearance and texture of the intertidal world.[3]

The aquatic realm has long stirred the human imagination. Today we know that life on this planet may have originated from primordial cells in ancient oceans. The protean sea also represents the cradle of trophic cascades science. Drawing from early ideas about predators, prey, and food webs, such as the work of Charles Darwin and Charles Elton, scientists began to ask bold questions. Some were framed ingenuously but nonetheless forged new theoretical ground. Yale University evolutionary ecologist G. Evelyn Hutchinson was among the young researchers asking questions. He simply wanted to know why there were so many species.

Homage to Santa Rosalia

Widely regarded a landmark paper in community ecology, Hutchinson's "Homage to Santa Rosalia, or Why Are There So Many Kinds of Animals?" inspired work by E. O. Wilson; Nelson Hairston, Frederick Smith, and Lawrence Slobodkin (HSS); Robert Paine; and others. It resulted from a 1950s trip to Sicily in which Hutchinson indulged his keen interest in entomology by collecting certain species of water bugs of the genus *Corixa*, which occurred in this region but had never been thoroughly studied.[4] In Palermo he visited the cave shrine of Santa Rosalia, an obscure twelfth-century saint. What he discovered in the spring-fed pool just below her sanctuary led him to dub her the patron saint of evolutionary studies. This was a simpler time in ecology, when scientists were still focusing on documenting species numbers and distribution. From the comfort of hindsight we can see that the world seemed to be moving far more slowly then, today's runaway global changes unimagined as yet. However, Hutchinson's observations in that grotto have been foundational to our understanding of how food webs function and how biodiversity may help create more resilient ecosystems in the face of global change.

So what did Hutchinson see? He found the water bugs he'd been searching for, of course, but their number and kind and behavior made him pause and wonder. He found two species of *Corixa*, which made him ask why not one, or twenty, or two hundred, for that matter? One of these beetle species was much larger than the other. In his ponderings about pond life Hutchinson evaluated Elton's food chain theory as the reason for the diversity he observed there. He examined Darwin's theory of natural selection as a factor affecting diversity. He also considered his student Robert MacArthur's nascent ideas about how a community increases in stability as the number of food web links increases. In that Italian grotto he further noticed more species of small and medium-sized organisms than large ones, similar to an observation Elton had made in the 1920s about community composition. This led Hutchinson to propose that so many animal species exist partly because a community with a complex trophic organization is more stable than a simple one. However, food chains' tendency to shorten as a result of unfavorable physical factors, such as lack of space, ultimately limits species numbers. Additionally he noted that larger-bodied species, although fewer in number, may exert a dominant influence on a system. As we shall see, the next generation of ecologists would proceed to argue Hutchinson's points and find other answers. But many of them cite "Homage to Santa

Rosalia" when asked what got them thinking about trophic relationships in communities.[5]

Uncharted Waters: Early Studies

Each type of landscape has its own parlance and terms used by those who study it. In the aquatic realm some of these terms are equally applicable to salt water and freshwater. Ecologists subdivide a body of water, be it an ocean or lake, into regions. Thus *pelagic* refers to open ocean, and *littoral* pertains to the shores of a lake, sea, or ocean; the *littoral zone* is another way of referring to the rocky intertidal zone. *Benthic* pertains to the bottom of a body of water, including sediment, and *benthos* refers to the organisms living in this region. In the marine world the *nearshore* or *neritic zone* extends from the low tide line to beyond the surf zone. The *longshore zone* refers to the area that runs farther out along the entire length of the ocean coast. Nearshore and longshore marine currents strongly influence intertidal ecology.

The rocky intertidal zone is the *ecotone*, or edge, that forms the permeable interface between sea and land. Particularly high in diversity and resources, ecotones represent a transition between two habitat types, with species often moving freely between them. A forest edge, where songbirds nest in the forest canopy but go out into adjacent grassland to forage for food, provides another example of an ecotone.

The crash of waves and twice daily advance and retreat of tidewater shape the rocky intertidal zone, one of the earth's most prolific habitats.[6] Upwelling of nutrient-rich nearshore water, which fertilizes plankton and plants, contributes to its abundance. The Pacific Northwest's temperate intertidal zone provided the setting for Robert Paine's revolutionary work on sea stars and mussels. In this research he conceptualized the food web interactions we have come to know as trophic cascades and defined their essential tenets and participants, such as keystone predators. Both ecologically and scientifically important, this landscape has inspired other studies that have changed the face of ecology.

In the early 1960s Paine began his job as a professor at the University of Washington, with National Science Foundation funding in hand. He proposed a study of the interaction between *Pisaster* and California mussels on the Washington coast but had trouble finding an appropriate research site. He finally found it in the spring of 1963 while on a field trip to the Olympic Peninsula with his marine invertebrate natural history students. He recalled the excitement he felt at the time: "I knew right away it was the system I'd been looking for. It had obvious pattern, because the organisms in that community created distinct horizontal bands. So it provided the ideal place to study predation, competition, and species diversity. My research assistant and I returned in late June after the quarter was over, found a suitable spot in Mukkaw Bay, and designed an experiment. That July we began removing *Pisaster* from our experimental plots."[7] As with Hutchinson's famed "Homage to Santa Rosalia," what followed has become enshrined in ecological history.

As we saw in chapter 2, Paine discovered that mussels were clearly affected by their primary predator, *Pisaster*, a species that generally lives lower in the intertidal zone and moves up to feed during high tide. Removing *Pisaster* caused the system to tip into a mussel-dominated state. As he kept up this removal over a three-year period, mussel distribution continued to expand and diversity of other intertidal species, such as barnacles, limpets, and chitons, continued to decline sharply. Meanwhile, in his control plot, which he left untouched, their distribution didn't change. Here the mussels formed a conspicuous band in the mid-intertidal zone, its position relatively stable, with species diversity maintained by *Pisaster* predation. Initially Paine's hypothesis was simply about competition and biodiversity, that is, how local species diversity is directly related to the efficiency with which predators prevent the monopolization of resources by one species. But in the process of doing this research he gained astonishing insights into how food webs work, especially the top-down effects of predation on community control. He published his findings in a 1966 article in which he linked diversity in intertidal rocky bottoms directly to predation intensity, although he found that other factors, such as disturbance (waves, storms), also influenced this system. In 1969 he introduced the term *keystone* to describe the ecosystem simplification that occurs when a top predator is removed.[8]

Paine soon added other field sites on the Olympic Peninsula, including Tatoosh Island, the largest of the islands lying off Cape Flattery, and broadened his investigation by adding sea urchins (*Strongylocentrotus* spp.) as a primary prey species. Now one of the most intensively studied sites in the world, Tatoosh Island has revealed how an interaction network links species and how local extinctions and environmental changes move through the food web. Ecological concepts explored here by Paine and his students include keystone effects, population dynamics involving body size, and impacts of environmental forces such as El Niño.

Next Paine expanded his work to an oceanic, intercontinental scale, working in sea star–mussel systems from Baja California to Chile to New Zealand. Again he found that the activities of a single keystone predator greatly modify a community's species composition and physical appearance. Where sea stars had been excluded mussels overtook the landscape, crowding out other species and eventually greatly reducing biodiversity. What he was finding, which the scientific world began to refer to as the "Paine effect," appeared to have widespread application to other marine systems. In a 1980 journal article about food web linkages and interaction strengths, he created the term *trophic cascades* to describe these top-down effects.[9]

A former student of Paine, Bruce Menge, has been investigating keystone predation with *Pisaster* and *Mytilus* in Oregon since the late 1980s. Partly in response to disagreement among ecologists about overly broad application of the *keystone* term, Menge took a close look at how a keystone functioned in his Oregon study areas and other rocky intertidal systems. Specifically he wanted to explore the interaction between bottom-up effects (disturbances such as wave action) and top-down effects (predation) to see whether *Pisaster* always performed as a keystone, and also to determine whether prey diversity increased as predation increased.

At Boiler Bay and Strawberry Hill on Oregon's central coast, Menge created experimental plots in which he and his students removed only *Pisaster* and left other predators, such as whelks (*Nucella lapillus*), intact. The results of this predator exclusion experiment suggested that *Pisaster* predation varied dramatically but, given time, was sufficient to eliminate mussels from the low intertidal zone and increase biodiversity. However, areas sheltered from waves in the upper intertidal zone experienced weaker and more variable predation. Here

Pisaster did not have such a strong keystone effect, and other factors, such as sand burial and low mussel recruitment (survival of juveniles to adulthood) had a stronger influence.[10] Menge's findings that *Pisaster* occurs as a keystone predator in some subhabitats but not in others raised questions about whether similar variation happens in other places—and to what extent keystone predation may be context dependent. Accordingly, he did a meta-analysis of studies in intertidal zones worldwide and found consistent keystone effects, which included an increase in prey biodiversity at sites with greater wave exposure.[11] This suggests that in this system, as in others, bottom-up and top-down forces interact, but keystone effects prevail.

Trophic Cascades in Kelp Forests

Over the years Paine has provided guidance to many students, including marine ecologist James Estes. Based at the University of California, Santa Cruz, the Estes lab has ongoing projects in the Aleutian Islands, California's central coast, the Channel Islands, Mexico, and Russia on how apex predators extend their influence to species other than their prey. Since the early 1970s Estes and his colleagues have studied sea otters, sea urchins, and kelp; in the mid-1990s they added killer whales. Although considerable research on marine trophic cascades has been done by others, here I focus on Estes' work, not only because of its significant contribution to science but also because of the valuable insights it yielded about what can occur when science challenges dominant resource management paradigms.

The sea otter–sea urchin–kelp cascade reported by Estes in Alaska takes place over a wide geographic area in the Aleutian Islands but represents a relatively simple system with just a few strongly interacting species. This archipelago comprises more than 300 mountainous islands, forming a rugged volcanic arc that stretches 1,100 miles southwest from the tip of the Alaska Peninsula to Russia. Originally populated by the indigenous Aleuts, this island chain forms the westernmost tip of the United States and the northernmost component of the Pacific Rim, creating a geographic barrier between the North Pacific Ocean and the Bering Sea. Danish explorer Vitus Bering visited this archipelago in

1741, initiating an era of exploitation by Russian trappers and traders, who harvested sea otter, seal, and fox furs and established settlements. The United States purchased the Aleutian Islands in 1867.

The sea otter is one of the largest members of Mustelidae, the weasel family. Once abundant from northern Japan to the Aleutian Islands and along the Pacific coast of North America south to Baja California, this species was extirpated by the late 1700s and nearly extinct by the early 1900s. In 1911 the North Pacific Fur Seal Treaty effectively protected the sea otter at a time when possibly as few as thirteen colonies remained. Its population increased slowly, to an estimated 150,000 otters by the early 1990s, living along rocky coasts, which provide sheltered feeding and resting areas.[12]

Sea otters are mammalian marine carnivores characterized by an ultra-thick coat of insulating fur, flipperlike hind feet, front paws with retractable claws, and a long, heavy body (40–100 pounds). They make undeniably charismatic poster animals for conservation; however, their button eyes and charming faces belie the fact that they are highly efficient carnivores. When they open their mouths to yawn, they reveal a set of teeth designed for business. Built for swift propulsion, they torpedo through the water, preying on benthic fauna such as sea urchins, sea stars, and abalone (*Haliotis* spp.). While otters can dive to 325 feet, they are seldom found in waters deeper than 150 feet and forage for food among nearshore kelp beds. Despite their hunting prowess, their thick pelts, which historically carried a high market value, and the fact that they are habitat specialists, adapted to live in only one type of environment, have made them vulnerable to extinction.[13] Killer whales (*Orcinus orca*) have recently become an additional threat to their conservation.

Sea urchins are hard-shelled marine animals distributed worldwide, ranging from the rocky intertidal zone to the deep ocean. They mainly eat kelp but occasionally eat meat and prey on one another. Their two primary predators include humans, who harvest them for commercial purposes, and sea otters. In the absence of sea otter predation they overbrowse kelp forests, denuding the ocean floor and forming urchin barrens, which contain a large number of urchins and little else. The marine equivalent of a desert, these barrens can cover several acres, containing thousands of sea urchins.[14]

Brown kelp (*Laminaria* spp. and *Agarum cribrosum*) belongs to the group of large, fleshy seaweeds that grow in cold, nutrient-rich waters off the coasts of several continents. Kelp attaches to rocky reefs via its holdfast—a rootlike knob that does not take in nutrients the way roots do in other types of plants. It forms multistoried forests with towering, cathedral-like canopies, their tops visible from the shore as they sway with the waves and ocean currents.[15]

A healthy kelp forest has a complex structure, like any terrestrial forest, with mid- and understory zones providing habitat for different organisms. Sunlight filtered through these giants' thick "leaves" nourishes microscopic plant plankton. Sea otters float on their backs at the ocean's surface, dozing, grooming themselves, and diving periodically to feed on the sea urchins and other creatures that live in understory thickets. Silver schools of fish dart through kelp tendrils, evading predators. While bottom-up factors, such as nutrient flow and water quality, influence this forest's growth, sea urchin herbivory represents one of the strongest factors determining its status from the top down within a trophic cascades framework.

In 1971 Estes was a PhD student newly arrived in the Aleutians, casting about for a research topic related to sea otter physiology. He hailed from a natural history wildlife background and was not an ecological theorist. He began working on Amchitka, a treeless island in the western archipelago subject to high winds and frequent, violent storms, which had a thriving otter population. As he was struggling to define his research project he had a catalytic chance meeting with Robert Paine, who was visiting the Aleutians to advise another student. Estes recalled that the discussion that inspired him to look at Shemya Island occurred in the dormitory before a movie screening. Although Estes was not familiar with HSS' green world hypothesis, all it took was a nudge from Paine to help him see beyond sea otter natural history to the ecological big picture of the processes at work in this system.

After listening to Estes' ideas, Paine said, "Have you thought about turning this around and looking at it from the perspective of what the otters are doing? I've been out here for a few days looking around, and I imagine they must be important consumers [predators] in this system. What if you visited islands without otters? What would you find there?"

Estes considered this and said, "You know, this would be an easy question to address because of the history of the species. We have this archipelago that was once completely populated by sea otters, and now they have recovered on a few of the islands but are still extinct on others. What more of a natural experiment could we ever ask for?"

"It's a great idea; do it," Paine said. He returned to Washington shortly thereafter, leaving Estes with much to think about.[16]

Estes soon visited Shemya, one of the Near Islands located far west of Amchitka, which sea otters had yet to recolonize. He was unprepared for the shock of what he encountered beneath the water's surface. He recalled:

> Most of my significant intellectual development probably happened in less than half a second, when I happened to see for the first time a system without otters. Ninety percent of my view of nature was codified right then, and the next forty years have been about quantifying that. When I made that first dive, I saw urchins all over everything, and no kelp. And I thought, wow, that must be it. We had a week at this island. I remember seeing it and wondering what we were going to do to chronicle this. And so in a short while I came up with a sampling procedure that could generate some data that would be descriptive of kelp and urchin abundance and population structure. After we sampled that island, we obtained comparable data from Amchitka.[17]

In areas without otters Estes found herbivorous sea urchins thickly carpeting the ocean floor—and no kelp. On reefs with otters he found a thick, green kelp forest and low numbers of sea urchins. The manuscript he and his co-researcher, John Palmisano, prepared was published in *Science* and provided empirical evidence of a trophic cascade (although Paine hadn't coined that term yet). Estes was still a graduate student and unaware of the future implications of his research.[18]

Estes worked in the Aleutian complex over the next four decades, in the Near Islands (the smallest and westernmost group in this chain) on Attu, which had a relatively new otter population; in the nearby Semichi Islands on otter-free Shemya; and in the Rat Islands (located sixty miles east) on Amchitka,

Figure 3.1. Kelp Forest in High Sea Otter Area

which had a carrying capacity otter population. Eventually he added the Andreanof Islands, which include Adak, another island with high otter numbers.[19] His objective was to measure a landscape-scale trophic cascade across the North Pacific in order to understand the processes that drive change in ecosystems.

Estes and his colleagues found many trophic connections in the nearshore system they studied. For example, they found that where a robust sea otter population existed, the accompanying reduction in sea urchins, a primary food source, caused glaucous-winged gulls (*Larus glaucescens*) to shift to a more varied diet that included fish.[20] Additionally, they linked otter effects to chemical defense compounds produced by kelp to make them less palatable to herbivores. Estes found concentrations of these chemicals (phlorotannins) to be ten times higher in Australian kelp forests than in those in the North Pacific, and he attributed this to a keystone species: the sea otter. Brown kelp has a nearly worldwide oceanic distribution in the nearshore zone; however, sea otters occur only in the Northern Hemisphere. In the absence of otters, which exert a controlling influence on urchins, southern kelp forests have had to evolve strong plant chemical defenses over millennia to thwart herbivory. Where sea otters have historically existed, kelp have had little need to develop such strong defenses.[21]

Other Estes lab research focused on how otter presence affected kelp species composition. Working in Torch Bay, Alaska, marine ecologist David Duggins removed sea urchins from experimental plots to simulate the effects of otters. In plots without sea urchins he found that the kelp species *Laminaria groenlandica* took over, outcompeting other kelp. On the basis of these findings Duggins suggested that sea otters, as keystone predators, clearly mediated major changes in kelp forest composition and that their near extinction must have created major ecological changes along the western coast of North America.[22]

Further deepening his lab's studies, Estes radio-tagged otters and followed their movements, to investigate changes in prey demographics and distribution related to patterns of predation and to look at the far-reaching effects of otter presence across multiple trophic levels. This was the sort of large-scale, well-funded research most scientists dream about doing. And then, in the early 1990s, the system began to change. Estes had no way of knowing the political consequences of the ecological changes he and his field crew began to observe.

Killer Whale Chain Reaction

It began in 1993 on Amchitka, with a pod of killer whales swimming near a kelp thicket inhabited by otters. Brian Hatfield, a member of Estes' research crew, observed a large male whale separating himself from the others and moving toward one of the otters. After an intense chase, the otter escaped by diving for safety. Hatfield had never seen anything like this in many hours of observation. That night he reported the incident to Estes, who was doubtful. However, evidence mounted over the next few years, making this phenomenon increasingly difficult to dismiss. In 1995 on Adak, four of Estes' field crew members observed a killer whale attacking and successfully killing an otter. One year later, also on Adak, Tim Tinker, Estes' research assistant, saw another killer whale take an otter.[23]

Details about otter distribution and demographics on Adak began to provide a plausible explanation. In Clam Lagoon, which had water too shallow for killer whales to enter, otter numbers remained high. In the adjacent open coastal environment, otter numbers had dropped exponentially. Estes published a paper about this in the journal *Science*, in which he posited that a collapse in populations of Steller sea lions (*Eumetopias jubatus*) and harbor seals (*Phoca vitulina*) across the western North Pacific had led to prey switching by killer whales. These two species had been key food sources for killer whales. He noted that sea lion declines began in the late 1970s, reaching a population nadir by the late 1980s, coinciding with the onset of the otter declines.[24] While this paper received substantial media attention, it did not appear to provoke a negative reaction in the marine science community. However, the suggestion that killer whales had switched prey to sea otters, and that this was related to the decline of other marine mammals, eventually would cause Estes to be strongly and unexpectedly challenged.

Over the next five years the otter decline continued at a slower pace. Alan Springer, a scientist from the University of Alaska, approached Estes with a hypothesis that played out across the entire North Pacific and involved extirpation of killer whale prey due to overharvesting by humans. As humans successively eliminated killer whale prey, these predators switched to progressively smaller food species, an effect like dominoes falling. The domino chain began like this: First killer whales preyed on great whales (e.g., gray, blue, humpback, right, sperm, and fin whales). When great whales became depleted, killer whales switched to *pinnipeds* (marine mammals with flippers, including seals and sea lions). When harbor seals declined, killer whales switched to Steller sea lions and, finally, to sea otters. By the late 1990s this prey switching, which Springer and Estes referred to in their sequential megafaunal collapse hypothesis, had created a four-level trophic cascade. Falling otter numbers provoked a surge in sea urchins, followed by excessive kelp consumption.[25] To better understand this domino effect, it helps to look at the natural history of killer whales.

Many people know killer whales as the appealing stars of marine theme parks, but these top predators have been known as "sea wolves," a name that carried negative connotations in earlier centuries. The killer whale, or orca, the largest member of the dolphin family, ranges worldwide from the Arctic to the Antarctic. Sleek bodied, with distinctive black-and-white markings and upright dorsal fins, they are highly efficient marine predators. While orcas have worldwide distribution, three Pacific populations exist: resident, offshore, and transient. Resident orcas, the most common, live in cohesive family groups (called pods) in North Pacific waters, such as Puget Sound in the northwestern United States. They feed primarily on fish and squid. Offshore orcas live in the deep ocean, often traveling in large groups of as many as sixty animals; we know little about their diets. Transient orcas range from the North Pacific to California coastal waters, roaming widely in small groups. Their diet consists primarily of large ocean mammals, including the great whales and pinnipeds.[26]

Transient killer whales, deprived of their historical food sources, began "feeding down the food chain," eating progressively smaller species as their prey became depleted as a result of overharvesting by humans. By the time Springer and Estes became aware of this megafaunal collapse, it was obvious that what they were observing was the end of the line. Their findings, published with several coauthors in 2003, sent a tsunami coursing through the fishing industry and government regulatory agencies.[27] Springer and Estes' research called into question the single-species strategy that had informed fishing regulations until then; their findings suggested that a much broader approach would be needed to protect biodiversity in marine systems. Further, if sea otters were to be protected via the Endangered Species Act, Springer and Estes' arguments that threats to their survival originated far up the food chain (i.e., with orcas) meant that recovery actions would have to reach across multiple geographic areas and food webs. In other words, the regulations and policies that had governed human activity in the sea until then were all subject to critical review.

Estes continued to study the otter decline. In 2005 he reported that by 2000, populations of this species at or near carrying capacity in 1965 had declined by 88 percent. Most local declines occurred from the late 1980s through the early 1990s. After that, otter populations stabilized at a level too low to keep sea urchins in check, maintained there by killer whale predation.[28]

Marine ecologist Terrie Williams' findings strengthened the link between killer whale predation, megafaunal collapse, and the otter decline. She investigated killer whales' metabolic needs and this species' potential impact on prey resources, including the missing sea otter biomass in the Aleutian Islands. To do this she compared killer whale daily caloric needs with changes in marine mammal populations. In her analysis she showed that fewer than forty killer whales could have caused the Steller sea lion decline and that the collapse of this top predator's prey base may have contributed to a sequential dietary switch, thereby initiating the declines Springer and Estes reported.[29]

Backlash

Three years after Springer and Estes presented their sequential megafaunal collapse hypothesis, several marine biologists launched a rebuttal. Published in *Marine Mammal Science* and other prestigious journals, these articles argued that killer whales did not eat great whales; that Springer and Estes' statistical analysis was erroneous; that the megafaunal collapse hadn't been sequential; that megafauna hadn't collapsed; and that otter deaths were due to disease or malnutrition.[30] In 2008 Springer and Estes published a response defending their work, which instigated another rebuttal. This point-counterpoint continued for several years.[31]

Over time Springer and Estes' work has been validated by studies conducted by Paine, Williams, and others. Paine chaired the National Research Council committee that investigated the Steller sea lion decline in Alaska waters. He and his colleagues found that from 1990 to the present, low sea lion harvest numbers suggested that top-down forces, such as the sequential megafaunal collapse, drove this population dynamic and presented the greatest recovery threat. Other scientists, such as Terrie Williams, helped craft the National Marine Fisheries Service's Steller Sea Lion Recovery Plan and provided further input about how this species' decline was connected to top-down effects.[32]

The sea otter decline has inspired further research from the Estes lab on how this species' presence or absence affects fish and the impacts this may have on other members of food webs. For example, working in nearshore marine communities in the Aleutian Islands, Shauna Reisewitz and her colleagues found that islands with high otter populations supported dense kelp forests, relatively few sea urchins, and abundant rock greenlings (*Hexagrammos lagocephalus*), a common kelp forest fish, while the opposite pattern (few greenlings, abundant urchins, denuded kelp forests) occurred at islands where otters were rare.[33]

In related research wildlife ecologist Robert Anthony and colleagues investigated the indirect effects of otter decline on another apex predator, the bald eagle (*Haliaeetus leucocephalus*). By comparing bald eagle demographics and diets before and after the sea otter decline, they tracked a domino effect that linked five species: sea otters, sea urchins, kelp, fish, and bald eagles. In areas with a high otter population in the 1970s and 1980s kelp flourished, sustaining high densities of rock greenlings, which provided food for eagles, gulls, and other carnivores. As otters declined, so did kelp beds, via the trophic cascade identified by Estes and colleagues. This triggered a corresponding decline in fish by eliminating their habitat, causing eagles, highly adaptable predators, to switch from eating a roughly even mix of fish, marine mammals, and seabirds to one dominated by seabirds (approximately 80 percent of prey consumed). Anthony found a positive correlation between this dietary switch and an increase in bald eagle reproduction rates, possibly driven by the higher nutritional value of waterfowl as compared with fish.[34] This study, with its unexpected finding, illustrates that we are only beginning to understand the complexity of indirect effects in food webs.

When it comes to resource management, it's still largely a bottom-up world. Government agencies, whether in the deep oceans or the northern Rocky Mountains, tend to focus on a species' conservation status, rather than taking an ecosystem approach, and seldom account for any but bottom-up environmental factors (e.g., food, stochastic effects such as climate). Springer and Estes' research suggested that this is inappropriate for marine systems because top-down effects caused the megafaunal collapse. Here, as in most systems, top-down forces interact with bottom-up forces in ecologically significant ways. Estes' and others' findings provide a scientific foundation for more progressive management strategies. With regard to top-down effects caused by sea otters, Estes says, "The point is not explaining that this is all that goes on in nature. The point is that this is a significant element of the dynamics of nature, and when you take the dominant predator out of most of these systems, it precipitates a ripple effect."[35]

As we have seen in relation to several issues in recent years, the interface between science, policy, and industry often leads to conflicts and controversies. Two recent examples from another field (forestry) include the debate over conservation of the northern spotted owl (*Strix occidentalis caurina*) and old-growth forests and the debate about postfire salvage logging involving Daniel Donato and colleagues, both of which took place in the Pacific Northwest.[36] While rigorous science can certainly be done on topics affected by resource extraction policy, sometimes science is used not to seek an empirical truth but to achieve political or industrial goals. The inherent danger is that in both cases researchers use the same vehicle to report their findings (publication in journals). This makes it difficult for people who rely on science to differentiate between an honest scientific debate and one that is malicious.

Trophic Dysfunction: Overfishing of Cod and the Great Sharks

Whether one looks at sea otters or cod, pelagic or nearshore zones, the Pacific or Atlantic, removal of apex predators sends powerful changes coursing through marine food webs.[37] University of Maine marine ecologist Robert Steneck used archaeological, historical, ecological, and fisheries data to identify three phase states in kelp forests in the Gulf of Maine, which extends 215 miles from Cape Cod to Cape Sable. Historically known for its abundant and diverse fish catch, this area includes the coastlines of New Hampshire, Maine, and Massachusetts north of Cape Cod. Sculpted by glacial activity, it contains a semi-enclosed gulf bounded by underwater banks.[38] As in other marine environments, fishing down the food web (which means harvesting progressively smaller fish species as the larger ones became scarce) caused the trophic collapse here.

The food web Steneck studied consisted of Atlantic cod (*Gadus morhua*), a keystone predator at the third trophic level, which limits green sea urchins (*Strongylocentrotus droebachiensis*) at the second level, which feed on kelp at the first level. Phase one, dominated by predatory fishes, occurred from 200 to more than 4,000 years ago and featured vast numbers of cod, low numbers of sea urchins, and multistoried kelp forests that supported a diverse assemblage of species. In the 1960s overfishing of cod and other predaceous fishes shifted this nearshore system into phase two, dominated by herbivorous sea urchins. It featured functionally nonexistent predators, urchin barrens, and overbrowsed kelp forests. Subsequent overharvesting of sea urchins tipped this system into phase three, dominated by predatory crabs. At first glance, in this last phase the system appears healthy again. Phase three again has luxuriant kelp forests, but it differs from phase one in that these forests are empty. These forests arose when sea urchin populations dropped below a threshold level because of unsustainable harvesting and their grazing pressure could no longer control kelp. In phase three Jonah crabs (*Cancer borealis*) assume the apex predator role, preventing sea urchins from recolonizing kelp forests. The resulting forest is thicker than in phase one and occupies a greater area but has far lower biodiversity.[39]

This community lost stability and capacity to recover from disturbances, a quality called *resilience*, at each phase when population densities of strongly interacting species fell below controlling thresholds. Thus decline in abundance of predatory cod in coastal waters destabilized phase one, releasing herbivorous sea urchins and leading to phase two. Sea urchin declines due to overfishing led to phase three, characterized by empty kelp forests. The rapid progression through the last two states—transitions referred to as *phase shifts*—suggests ecosystem instability. The resulting ecosystems are said to be in an *alternative stable state* because these phases are capable of persisting. As anthropogenic changes progress in ocean communities, low-diversity ecosystems such as this are becoming more common.[40]

Farther south along the Atlantic seaboard sharks have fared the same as cod, leading to similarly destabilized ecosystems. Long the subject of legend, the great sharks are top predators feared by their marine prey and humans alike. Late marine ecologist Ransom Meyers investigated how in recent decades they have been overfished from the Atlantic because of demand for their fins and meat.[41] Between 1972 and 2007, twelve of fourteen shark species experienced significant decreases in the mid-Atlantic. For example, sandbar sharks (*Carcharhinus plumbeus*) declined by 87 percent, blacktip sharks (*C. limbatus*) by 93 percent, and tiger sharks (*Galeocerdo cuvier*) by 99 percent. These sharks kept cownose ray (*Rhinoptera bonasus*) numbers low. Fewer sharks resulted in a twentyfold increase in this mesopredator, which consumes large quantities of bivalves (organisms encased by a hinged shell).[42] Marine ecologist Charles Peterson discovered that by 2004 cownose rays had nearly extirpated scallops from

Chesapeake Bay to South Carolina's Cape Lookout, effectively eliminating commercial shellfisheries.[43]

This ecological collapse compounded the harmful effects of disease, overharvesting, habitat destruction and fragmentation, and pollution, which already had reduced bivalve populations. The resulting cascade can potentially extend to the sea grass community in estuaries and shallow nearshore waters. Sea grass meadows provide critical habitat for juvenile fishes and lobsters; their potential loss carries the threat of broader ecosystem degradation.[44]

Trophic Cascades in Lakes: Long-Term Ecological Research

Ten thousand years ago retreating glaciers created literally thousands of lakes in the northern United States from the upper Midwest to the northeast. Today when you fly over these areas the lakes look like bright jewels set in a rich green matrix of forest and farmland. But from the ground it becomes apparent that many are not doing well, filled with algae, their water murky and polluted by human land use. Activities that take place miles away can indirectly affect lake ecology via runoff and nutrient flow. Intimately connected to the watersheds in which they are embedded, lakes freely exchange nutrients and organic matter with land through the downslope movement of water. This means that a combination of top-down and bottom-up forces influence lake ecosystems. For example, manipulating fish harvest sends energy rippling through food webs from the top down.[45] However, agriculture and urbanization create strong disturbances that make nutrients such as nitrogen and phosphorus flow into the water, affecting the system from the bottom up. Optimal lake productivity and health depend on moderate levels of nutrients, especially phosphorus. As we shall see, having too many nutrients can be disastrous to lake health.[46]

Since the 1920s scientists who study lakes, called *limnologists*, have been concerned with a condition called *eutrophication* wherein too many plants growing in the water can make lakes die. Caused by excessive nutrients, such as phosphorus, this process features progressive invasion by vegetation, particularly blue-green algae (cyanobacteria), which is toxic to humans and other animals in high concentrations and makes lake water cloudy. During the latter stages of eutrophication overabundant plant life chokes off the body of water by consuming most of the oxygen.

John Brooks and Stanley Dodson were among the earliest to map community structure in lake ecosystems in the 1960s. These food webs have four principal levels: carnivorous fish (also called piscivores) at the top, or fourth, level; plankton-eating fish (planktivores) at the third level; *zooplankton*—free-floating organisms, mainly crustaceans and fish larvae, characterized by involuntary movement—at the second level; and *phytoplankton*, tiny photosynthetic organisms, at the first level. Piscivores eat planktivores, which eat zooplankton, which eat phytoplankton. Brooks and Dodson did their landmark research in Branford, Connecticut, at Cedar and Linsley ponds, focusing on competition between various species and sizes of zooplankton, including herring (*Alosa aestivalis*) larvae, a dominant predator. They found larger zooplankton to be more efficient at obtaining food. This research into the intricacies of competition in this system and the competitive edge that results from larger body size built on Hutchinson's early work.[47]

Stephen Carpenter, a limnologist from the University of Wisconsin, was inspired by Brooks and Dodson's work to look at the whole food web in lakes and how phosphorus inputs might influence it. Here I focus on his early work because he was the first to quantify trophic cascades in lakes. He did much of this work in northern Wisconsin, at the University of Notre Dame Environmental Research Center. Established in 1940, this facility contains thirty lakes, including whole watersheds, and provides a unique opportunity to learn about aquatic ecosystems not subject to disturbance by the public (e.g., fishing and agriculture). By the time Carpenter began his research in the 1980s, scientists had learned that phosphorus inputs could account for less than 50 percent of the energy flow in lakes, via production of phytoplankton. This unexplained variation intrigued him. Further influenced by Paine's work, he tested a trophic cascades hypothesis to explain differences in productivity among lakes with similar nutrient supplies but contrasting food webs.[48]

The food web Carpenter studied consists of the four levels identified by Brooks and Dodson. In this system one can reduce the amount of blue-green algae in a lake that is becoming eutrophic by adding piscivorous fish, such as largemouth bass (*Micropterus salmoides*) or northern pike (*Esox lucius*). Altering lake food webs in this manner is called *biomanipulation*. The resulting ecological cascade works like this: bass and pike reduce numbers of planktivorous fish such as yellow perch (*Perca flavescens*), which in turn increases zooplankton (*Daphnia* sp.), which eat phytoplankton, thereby improving water quality. When humans overharvest carnivorous fishes, this chain reaction works in reverse: a decrease in piscivores creates an increase in planktivores, a corresponding decrease in zooplankton, and an increase in phytoplankton. However, Carpenter quickly found this framework inadequate to fully describe the complex processes at work in lakes, such as changes in fish diet and habitat, zooplankton nutrient recycling, and, perhaps most astonishing, changes in fish foraging behavior in response to predation risk.[49]

In his initial experiments Carpenter biomanipulated three lakes at the Notre Dame Environmental Research Center: Paul, Tuesday, and Peter lakes. In Paul Lake, which he used as a reference, or *control*, site, he did nothing and observed productivity, which varied from year to year because of *abiotic* (nonliving, bottom-up) factors such as climate. This enabled him to measure biomanipulation effects at the other lakes. He added piscivorous fishes to Tuesday Lake, which reduced planktivorous fishes, increased zooplankton, and reduced algae. In Peter Lake he tried the opposite. He reduced piscivores, which caused planktivorous fishes to increase, zooplankton to decrease, and algae to increase. However, he found surprising results in this lake that had to do with the ecology of fear. According to Carpenter,

> When 90 percent of the largemouth bass were removed from Peter Lake, and 49,601 zooplanktivorous minnows added shortly thereafter, the minnows behaved as expected and immediately began exploiting the large zooplankton as prey. That lasted about two weeks. Perception of predation risk owing to the remaining bass population rose, and by the end of the first month nearly all of the minnows were densely aggregated in refugia (beaver channels), where they gradually starved and many were eaten by piscivorous birds. That result was unexpected and our monitoring program represented it only sparingly.[50]

Carpenter continued his research over the next two decades. In retrospect he emphasizes the importance of long-term studies that encompass whole lakes and watersheds. He broadened his work to include Lake Mendota in Madison, Wisconsin, and the US Forest Service's North Temperate Lakes Long-Term Ecological Research site in northern Wisconsin, which comprises seven lakes linked through a common groundwater system. His Lake Mendota work, which we'll take up in chapter 7, shows the ecological restoration potential of biomanipulating lakes using trophic cascades principles.

Riverine Cascades

Mary Power, an aquatic ecologist from the University of California, Berkeley, set out to look for trophic cascades in streams, beginning in the early 1980s as Robert Paine's graduate student. In Panama she studied the algae-eating armored catfish (Loricariidae family) and found its density negatively related to the amount of forest canopy over streams, because the canopy impeded algae growth. She followed this research with postdoctoral work in the midwestern United States on riparian food webs, and from there she went on to do seminal work in northern California on the Eel River.

The Eel River rushes out of northern California's rough coastal mountains, flowing northwest and then west through Mendocino National Forest as it winds through narrow canyons and ancient redwoods and past small mountain communities on its way to the ocean. Although its headwaters were dammed nearly a century ago, in 1981 it received federal wild river designation to help prevent further damming and protect its legendary steelhead runs. Halfway to the ocean four major tributaries join the main channel, widening the river and creating a gently sloping valley. The Eel empties into the Pacific Ocean 200 miles from its source, near the town of Eureka, amid coastal redwoods. Powers set her research on this river in a six-mile reach of one of its tributaries, the South Fork, which forms part of the University of California's Angelo Coast Range Reserve. Here she discovered yet another example of top-down ecosystem regulation.[51]

Streams have more complex food webs than many other ecosystem types, with an intricate interaction of multiple species at each trophic level. However, the fundamental top-down and bottom-up dynamics here resemble those in simpler systems. Scientists have widely acknowledged the importance of riparian vegetation, which regulates the amount of light that reaches streams and controls water temperature. It also provides habitat and nutrients for aquatic and terrestrial organisms and helps maintain energy flow via photosynthesis and nitrogen fixation. As in other systems, trophic cascades have a powerful effect on vegetation.[52]

To gain a better understanding of how riparian food webs function, it helps to identify the strongly interacting species in this system. Midges, also referred to as chironomids—small wormlike organisms that develop in lakes, ponds, and streams—are important participants in this food web. Often the most abundant aquatic insects in these habitats, they feed on algae and other organic matter and are preyed upon by fish. Also key to riparian food webs, the filamentous green alga *Cladophora glomerata* provides nourishment and habitat for many species. Its floating matlike structures create physical variation within streams, forming turfs that lodge in rocky riffles or areas of slow water and dominate stream biomass during spring and early summer. When released from predation, dense infestations of chironomids weave algae into tufts, reducing algal cover and biomass.

The food web Power and colleagues studied has steelhead (*Oncorhynchus mykiss*) and California roach (*Hesperoleucus symmetricus*) at the top, or fourth, trophic level. These species feed on predatory insects and fish fry, as well as on three-spined stickleback (*Gasterosteus aculeatus*), a secondary predator at the third level. Stickleback feed on chironomid larvae, at the second level, which feed on *Cladophora* at the first level. Power experimentally manipulated fish populations to test ecologists Stephen Fretwell and Lauri Oksanen's hypothesis that food webs with an odd number of levels will be "green," characterized by top-down energy flow, while those with an even number will be "brown," characterized by bottom-up energy flow and overbrowsed vegetation.[53]

By constructing twelve large cages over rocks that supported large standing crops of algae, she was able to see how the food web developed with and without top carnivores. The cages had mesh walls that were permeable to riparian insects and small fish but not to top predatory fish. In *enclosures*, which represented a four-level food web, she live-captured and removed preexisting fish and then restocked the cages with steelhead and roach to reflect the proportions and density existing in the open river. In *exclosures*, which represented a three-level food web, Powers kept out these predators.

After just five weeks Power found striking differences between the algae in enclosures and exclosures. In the enclosures (four-level food web), which contained top predators, chironomid infestations had significantly reduced algae. In contrast, exclosures, which represented a three-level food web, contained lush mats of *Cladophora*. Here small fish suppressed chironomids, enabling algae to grow. Power's findings supported her prediction that algae in a three-trophic-level web will thrive, while algae in a four-level web will become overbrowsed and reduced to relatively barren levels.[54]

After her work on the Eel River, Power went on to complete more than two decades of research on riparian food webs worldwide, focusing on context dependency of the interactions she observed. She and her colleagues found that while top-down effects occur in riparian systems everywhere, climatic (bottom-up) effects, such as floods and drought, can weaken them.[55] Over the years Powers' diverse research has probed riparian food web complexity and helped researchers in another, equally complex type of aquatic ecosystem better understand trophic interactions.

Trophic Trickles: Life in the Coral Reef

Tropical nearshore environments include coral reefs and sea grass meadows, historically populated by sea turtles (*Chelonia mydas*), manatees (*Trichechus* spp.), crocodiles (*Crocodylus* spp.), sharks, and other apex predators. As in all landscapes, the trophic dynamics here were once tightly calibrated, containing organisms that had coevolved. As a result of human exploitation, pollution, and climate change, many of these species are now gone or reduced to the point where they are no longer ecologically effective, breaking historical trophic linkages. Additionally, fishing down the food web has skewed fish abundance, biomass, and community structure to favor smaller-bodied, herbivorous fishes. This can have profound implications on energy flow in an ecosystem.[56]

Coral reefs provide the foundation for marine life and support millions of humans globally. Considered the "rain forests of the ocean," they harbor highly diverse flora and fauna, as many as 2 million species by some estimates, including 25 percent of marine fish species.[57] Reefs come in different sizes and shapes, sculpted by ocean currents and landforms. They grow from the bottom of the ocean and rise upward and out of the water, in the process making shallow lagoons on their tablelike surfaces. Colonies of tiny coral polyps, taxonomically related to jellyfish, create reefs by removing calcium from ocean water and secreting a calcium carbonate (limestone) cuplike outer skeleton called an *exoskeleton*. Their fleshy bodies have tendril-like tentacles that they extend to feed, capturing crustaceans and other small animals. When finished eating, they withdraw the tentacles into the shelter of their exoskeleton. Within their soft bodies polyps harbor a single-celled brown alga with which they have a symbiotic relationship. Via photosynthesis this alga helps polyps produce calcium carbonate. As polyps grow and bud on top of one another, they create the lacy geometric pattern that results in a coral reef. When they die they leave their exoskeletons behind, allowing new polyps to develop on their remains. Some of the reefs thus formed may be thousands of years old. Corals require specific habitat to thrive, such as warm, clean salt water penetrated by sunlight; too much sediment can smother polyps, and too much freshwater can kill them. Most shallow water reefs occur between 30 degrees north and south of the equator, where the water is the right temperature.

Coral reefs are among the earth's most endangered habitats, with 10 percent lost already and 70 percent due to die within twenty to forty years unless we significantly reduce ocean pollution and halt fishing practices lethal to them (e.g., trawling, fishing with explosives).[58] Marine reserves, used worldwide to prevent overfishing and protect biodiversity, are proving useful in this effort. Varying in size and design, these reserves typically include fishing bans, with buffer zones between protected areas and ocean regions open to harvest. This strategy is particularly effective for conserving threatened populations of large, carnivorous fishes. Although no marine system can be considered "closed," reserves provide a relatively controlled environment, so they make excellent sites for trophic cascades experiments.

Evidence of cascades is sometimes equivocal in coral reefs because of their high biodiversity, which creates trophic flexibility and redundancy. Returning top predators by protecting them from fishing can generally restore these systems; however, the effects are not always straightforward because of complex interactions that have been severed or modified by fishing. For example, in the Bahamas' Exuma Cays Land and Sea Park, one of the largest and most successful marine reserves, a fishing ban produced surprising results. When the reserve was established, piscivorous fishes, such as the Nassau grouper (*Epinephelus striatus*), predictably increased. Scientists anticipated that in this four-level food web the ban would decrease herbivorous fishes at the third trophic level, via predation, thereby diminishing grazing on kelp that grow on coral reefs. In temperate oceans, kelp, also called *macroalgae*, or large, fleshy algae, benefit ecosystems, but in the tropics too much of it can threaten corals by competing with them for space on reefs. As predicted, the fishing ban increased carnivorous fish populations, but it unexpectedly caused large-bodied grazing parrotfish (*Cheilinus digrammus*) to increase. No longer harvested by humans, they escaped predation by other fishes because of their size. Parrotfish eat kelp; as they increased, kelp declined, benefiting corals via a trophic cascade. Thus, in this four-level food web, more predaceous fishes failed to produce the expected outcome: fewer parrotfishes and more kelp.[59]

Sometimes, despite these systems' intricate food webs, overharvesting by humans can rapidly provoke an ecological meltdown. For example, in Jamaican coral reefs, overfishing during the 1960s and 1970s extirpated carnivorous and herbivorous fishes, causing the sea urchin *Diadema antillarum* to irrupt and emerge as the dominant grazer. This meant that this food web went from having multiple species of predators and grazers to a single grazer. And then, in 1983, disaster struck when an unknown disease killed more than 98 percent of the sea urchins. Within weeks of the die-off, large, fleshy algae exploded, in two years increasing from 10 percent to 90 percent of the vegetation cover, presenting a serious threat to coral recruitment. The moral here is that trophic complexity, as in coral reefs, can provide a false sense of security, and even the most diverse systems can quickly lose species and stability when subjected to severe anthropogenic modification. Additionally, simplification of trophic pathways can lead to dynamics (e.g., sea urchin irruption) that increase the spread of pathogens, which can have devastating ecological effects.[60]

IN THIS chapter we have explored trophic cascades in aquatic systems from the North Atlantic to California rivers to tropical oceans. While the same mechanisms operate in all these systems, high biodiversity in rivers and coral reefs creates complex food webs in which cascades can take surprising pathways. Nevertheless, studies show that as human modification of ecosystems continues apace, characterized by overharvesting of aquatic resources, we will be seeing more of the food web instability identified by the researchers whose work is profiled in this chapter. The megafaunal collapse in the North Pacific reported by Springer and Estes provides a compelling case study that illustrates the challenge of getting managers and government agencies to acknowledge the relevance of keystone predation.

CHAPTER FOUR

Why the Earth Is Green: Terrestrial Cascades

Debate has always been a fundamental part of how science works. Beyond merely being philosophical discourses about the nature of the world, arguments advance and shape science. The scientific method begins by asking questions and then trying to find the answers, each answer begetting more questions. And so it was with the green world hypothesis, one of the most contested topics in ecology since the 1960s. A leading argument focused on whether terrestrial communities have fundamentally different dynamics from aquatic communities. In the 1990s ecologist Donald Strong and others suggested that trophic cascades may be "more wet than dry" and may not occur in terrestrial systems. Getting to the root of this was hampered by the almost insurmountable logistical challenge of applying the experimental approach used in aquatic systems to large terrestrial mammals. Additionally, there was a matter of scale. All aquatic systems, except for the deep oceans, tend to have fairly compressed scales compared with those of terrestrial communities. Marine ecologist Bruce

Menge suggests that terrestrial systems would show, if we could study them at the proper scales, the same range of trophic dynamics as aquatic systems.[1]

A landmark meta-analysis by Yale University ecologist Oswald Schmitz in 2000 helped shed light on this matter. Looking at forty-one studies in various terrestrial systems, he found trophic cascades common. Exceptions occurred when plants had antiherbivore defenses or when herbivore diversity was high. Strength of top-down effects varied with the type of carnivore and plant response mechanism measured, with vertebrate carnivores having stronger effects than invertebrates.[2] Schmitz' work opened the floodgates for trophic cascades research in additional terrestrial systems, most notably in those harboring large mammals. Many questions remain, including what direction researchers should take to deepen our understanding of trophic cascades, which are anything but simple, characterized as they are by interplay of top-down and bottom-up energy. Scientists have an incomplete understanding of predation risk mechanisms, which lie at the heart of trophic cascades studies of large mammals, such as elk (*Cervus elaphus*). Spirited debates about these topics have been ongoing since the mid-1990s. In this chapter I explore these concepts and present various researchers' perspectives, using Isle Royale and Yellowstone national parks to illustrate how top-down and bottom-up forces intertwine and the different ways predation risk may operate.

Trophic Architecture

In any discussion about terrestrial trophic cascades, it's helpful to become acquainted with some of the terms and methods used in this work. Ecologists evaluate a variety of trophic mechanisms and plant responses. These responses leave signatures, or architectural patterns, on a landscape, measured using some of the tools discussed here.

Trophic cascades originate in the mechanism of disturbance. A disturbance is any event that disrupts an ecosystem and changes nutrient flow or the structure and condition of the physical environment.[3] Many types of disturbance exist in the natural world, such as storms and fire. Two types of disturbance characterize trophic cascades: predation and herbivory. Predation is a deceptive force because in a community the act of a predator killing an herbivore can seem like a small, discrete event. But predation adds up and can be measured by changes in prey population and behavior, as discussed in chapter 2. On longer time scales it drives evolution as a selective force. Herbivory leaves sharp marks on a landscape. One of the most telling, the *recruitment gap*, involves the missing tree age classes caused by chronic herbivory. A healthy tree community has an age distribution shaped like a reverse letter J. Such communities contain a vast majority of young trees, with progressively fewer trees in older age classes. With chronic herbivory young trees, such as aspens (*Populus tremuloides*), cottonwoods (*Populus* sp.), and balsam firs (*Abies balsamea*), often cannot grow above browse height—the height to which an ungulate can eat.

As soon as a sapling puts on growth, the tender current-year growth tip at the end of its primary stem, called the *apical bud*, becomes vulnerable to ungulates. A favorite food for them, the bud provides a strong nutrient punch, similar to what dark, leafy greens do for humans. Where ungulates have irrupted, this leads to age distributions that look like the letter U: lots of young trees below browse height and many old, dying trees, but essentially no middle-aged ones. In this scenario, seen worldwide today, within a few decades entire forests will

Figure 4.1. Aspen Recruitment Gap, Glacier National Park

die out unless something changes to enable young trees to recruit into adults. That something, according to many of the researchers profiled in this chapter, may involve wolves.

The best way to measure a recruitment gap is by using *dendrochronology*—the study of tree rings. To do this ecologists core trees, count the rings in the cores, and measure the trees' diameter. They determine the relative age of trees in a stand by taking a representative sample and creating a mathematical relationship between age and diameter. This can be used to pinpoint when trees stopped recruiting into adults. A *release* results when herbivory pressure declines with the return of keystone predators, or it can be caused by increased resources available to trees. When caused by return of predators, this is driven by a reduction in herbivores (a density-mediated response) or a change in their behavior (a behavior-mediated response), as when ungulates avoid foraging in risky areas. As noted in chapter 2, the shrubs growing taller on my land after wolves returned are an example of how predator presence may be linked to a release from herbivory.

In addition to recruitment gaps and releases, the way young trees grow provides another bold signature. Many species of deciduous trees are genetically programmed to grow straight and tall, with alternating branches. While climate and fire can affect growth, disease and herbivory can also make a tree deviate from this pattern. Marks left by disease, which include blackened, atrophied stems, are easily distinguished from those caused by browsing. When an ungulate browses an aspen, it nips off the apical bud. This causes the stem to branch or grow crookedly to one side, leaving a kink. With successive browsing over the years, the stem begins to zigzag. If browsed enough times, eventually a tree takes on a shrublike, bonsai appearance and dies from chronic herbivory. Where a tree stops growing in any given year, regardless of whether it is browsed or not, it forms an apical bud scar, which appears as a faint ring going all the way around the trunk. You can determine a young tree's age by counting its apical bud scars. Defined as the aboveground structure of a plant, *plant architecture* can help determine the plant's height in previous years as well as its current and historical browsing. Ecologists measure browsing intensity by studying the zigzag pattern of stem growth and counting the apical bud scars.[4] Plant architecture reflects the ecological forces at work on trees and can be used to find a correlation between predator presence and absence. But these effects are not limited to large ungulate herbivores. Other species, such as monkeys and iguanas, can be powerful trophic architects, redesigning ecosystems from the top down by influencing how things grow.

Island Cascades: Barro Colorado Island and Lago Guri

John Terborgh's visionary research in tropical island communities has helped us understand what causes ecosystem simplification. His early work took place on Barro Colorado Island in Panama, which was created in 1923 when the canal was cut, flooding local areas. Formerly a forested hilltop, this new island became a Smithsonian Institution nature preserve and research site. Not long after its creation, Barro Colorado lost its large carnivores, beginning with the puma (*Puma concolor*). The jaguar (*Panthera onca*) and harpy eagle (*Harpia harpyja*) followed, allowing mesopredators to flourish, some of them seed eaters, such as the peccary (*Catagonus wagneri*), and some nest raiders, such as the opossum (*Didelphis* spp.). By 1970 forty-five species of birds had gone extinct, many of them ground dwellers. Terborgh and Blair Winter were among the first to find the reason: mesopredator release. Terborgh's next research site would provide even more dramatic evidence of the ecological changes that occur when predators are excluded.[5]

A natural experiment arose in 1986 when the expansive Guri Dam hydroelectric reservoir flooded the Caroní River valley in Venezuela. The flood covered 1,660 square miles, fragmenting the formerly continuous dry mainland forest, creating hundreds of islands ranging in area from a quarter acre to one and a half square miles. As had Barro Colorado, some of the smaller islands quickly lost big things: jaguars and harpy eagles.[6] This created an opportunity for predator exclusion research, which Terborgh and colleagues believed could provide a microcosm of what is happening globally where large predators are confined to a fraction of their former ranges.

As the water rose, wildlife sought higher ground and became concentrated on small habitat patches. Terborgh's first survey, in 1990, suggested that more than three-quarters of all vertebrates present on the mainland had already disappeared from the small islands, leaving imbalanced animal communities. He and his colleagues found pollinators and seed dispersers underrepresented, with large predators entirely missing. On the other hand, nearly all animal species that persisted had population densities twenty to thirty times above normal on the mainland. Hyperabundant groups included birds, some lizards and amphibians, spiders, small rodents, and generalist herbivores such as common iguanas (*Iguana iguana*), red howler monkeys (*Alouatta seniculus*), and leaf-cutter ants (*Atta* spp., *Acromyrmex* spp.).[7] At such high densities these herbivores left much larger ecological footprints. Among them, the howler monkey was most emblematic of the ensuing ecological degradation.

Terborgh worked on fourteen islands. On small islands, birds, lizards, amphibians, and spiders preyed on invertebrates. Medium-sized islands supported all predators found on small islands plus a mesopredator, the capuchin monkey (*Cebus olivaceus*); a rodent, the agouti (*Dasyprocta* spp.); and an armadillo (*Dasypus novemcinctus*). Two large islands and two mainland sites served as controls containing all primates in the region, mesopredators, and large predators. Herbivore abundance varied inversely with island size, with small islands supporting the most herbivores.

Terborgh hypothesized that altered animal communities on small and medium islands would create a trophic cascade. Returning to Lago Guri yearly, he closely monitored vegetation change and found low recruitment of saplings on islands lacking large predators. Leaf-cutter ants emerged as the dominant herbivore, helped in the canopy by hyperabundant howler monkeys and common iguanas. Unchecked by predators, the ants devoured the vegetation and left little more than barren red earth and aberrant thickets of thorny vines. On small and medium islands, woody plant mortality exceeded recruitment for nearly all species. This imbalance led plant communities to collapse and be replaced by new ones not found on the mainland. Terborgh found scale the single most important factor influencing predator loss—the larger the island, the greater its ability to support a diversity of predators.[8]

Beneath these ecological signatures lay a powerful lesson. One would expect howler monkeys liberated from predation to enjoy a more peaceful existence.

Researcher Gabriela Orihuela found social dysfunction instead. Hyperabundant howlers came together in social groups less and played less. As they denuded their habitat by overbrowsing, the trees began producing noxious defense compounds, which made the howlers sick. They would eat until they were full and then vomit. Liberated from top-down control, howlers experienced a far more challenging world.[9]

I asked Terborgh if there had been a particular moment afield when his understanding of island systems had coalesced. He spoke of a time when he was studying birds on the little Guri islands. As he worked, he noticed significant differences between otherwise nearly identical islands—all of them little islands about one and a half acres in size. Some had practically no birds, while others had throngs of them, and this puzzled him. "One morning, before getting up, I lay there listening to birdsong, wondering what was going on with these islands. Then it occurred to me that the islands that had lots of birds all had howler monkeys, and those that had few or no birds didn't have any. So the difference seemed to correlate with the presence or absence of howler monkeys. And of course how one gets from howler monkeys to birds is a long story, but it has to do with trophic cascades."[10]

Terborgh expected this downward ecological spiral to run its course in a few decades, leading to considerable biodiversity loss. However, his Lago Guri work ended in 2003, when a prolonged drought reconnected some of the islands, enabling monkeys and other wildlife isolated by the water to flee. All the same, this research, conducted from 1993 through 2003, helped establish top predators as indicators of ecosystem health. Predator range contractions worldwide suggest insidious ecological imbalances. By the time managers and others become aware of and acknowledge these changes, it may be too late to reverse changes in plant communities. Terborgh's work shows how awareness of the ecological damage caused by removal of keystone predators can help us identify warning signals of impending extinctions.[11] And Orihuela's and Terborgh's examples might make one wonder how many other mysterious conservation problems, such as amphibian declines, have causes related to trophic cascades.

Evolutionary ecologist Jared Diamond points out that while Terborgh's Lago Guri experiment is ostensibly about control of prey by their predators and of plants by their herbivores, there is far more to it than that. He likens this research site to a kaleidoscope that gets shaken up hundreds of times, with some general tendencies emerging among the resulting patterns but with fascinating differences because of chance, yielding plant and animal communities that would never exist normally.[12]

Islands provide fascinating places to study predator-prey interactions and look closely at the top-down and bottom-up dance. Work done on a famous North American island has yielded vital insights into these relationships.

Isle Royale Cascades

The story of trophic cascades sciences related to the wolf (*Canis lupus*) begins with Rolf Peterson's seminal work on Isle Royale. I met with him at his Michigan Technological University lab, where he spends the small amount of time he is not in the field. Officially retired since 2007, Peterson says that the main change in his life is that now he is able to spend more time on the island, continuing work begun in the 1960s, when he was a graduate student of early wolf biologist Durward Allen. A soft-spoken, modest man, Peterson has a diffident manner that belies the fact that he is one of the most distinguished scientists studying predator-prey interactions today. His science has blazed a trail for researchers working on relationships between wolves and their prey.

Located in Lake Superior, Isle Royale National Park is fifty miles long and approximately eight miles wide. Dense coniferous and hardwood forest covers this rock-ribbed, glacier-carved island, which was born with the recession of the last glacier, some 10,000 years ago. The Ojibway Indians called it Minong—a name that carried several meanings: turtle, the high place, place of the berries.[13] Today glacial till and basalt boulders soften its contours, and the island lies there in all its raw, elemental splendor. Greenstone Ridge, which provides the island's backbone, offers superb views. Occasional openings in the forest canopy reveal grassy meadows, chains of beaver ponds, and many inland lakes. This inscrutable wilderness holds many ecological secrets, which it gives up slowly.

I visited Isle Royale in 2005 when I was about to begin my PhD studies. Although I ended up doing trophic cascades research closer to home, Isle Royale's stark beauty and scientific potential captivated me, as it had others before me. It's not a place for the fainthearted. Lake Superior's icy waters effectively isolate this long green jewel of an island. Out of the ordinary run of human lives, its rocky terrain, extreme weather, and complete roadlessness make it the least visited of the national parks. Paradoxically, it has been one of the most studied parks, beginning in 1904, with the Charles Adams expedition's biological inventory, and continuing with wildlife biologist Adolph Murie's 1930 study of moose (*Alces alces*) and vegetation.[14] Regardless of the hazards, wonders await those who journey across the rough lake waters, shoulder their heavy packs, and place their feet upon the rocky trails that penetrate the island's silent green interior.

The nineteenth century brought intensive human activity to Isle Royale, with attendant guns and other weapons of harvest. Humans had extirpated beavers and caribou by the late 1800s. At the beginning of the twentieth century, moose colonized the island, probably by swimming across the lake. By 1930 Murie noted that the moose population had exploded, overbrowsing the island. A breeding population of wolves became established through dispersal from the nearby Ontario mainland in the late 1940s. Wolves are the sole moose predator here. The only ungulate species on this island, moose provide 90 percent of the wolves' diet. Isle Royale was considered an ideal wolf sanctuary, because of the abundant prey, and thus a good place to study them.

Allen began his work here in 1958. When Peterson joined him, wolf and moose populations seemed stable, suggesting that predation was regulating moose density below the level at which food might be limiting. The population cycling that followed illustrates the complexity of top-down and bottom-up effects but supports the concept of the wolf as a keystone species. Between 1959 and 1970 wolf numbers hovered in the mid-twenties. Moose stayed around 700 until 1967, when their numbers climbed, reaching an all-time high of 1,500 by 1973. They overbrowsed the island's vegetation, particularly balsam fir, a key winter food. After a lag of a few years, wolf numbers started to rise, reflecting the increase in prey abundance. Malnutrition combined with several severe winters caused the moose population to plummet to 900 in 1980, while wolves soared to a high of 50, crumbling the notion that wolf-moose relationships were relatively static and making it clear that this research had to continue.[15] While bottom-up effects (winter severity, which limits food availability) prompted moose decline, wolf predation (top-down influence) accelerated it. In the late 1970s Peterson left for Alaska for three years. What he found upon his return inspired his trophic cascades research with Brian McLaren:

> I came back to our base camp at Bangsund Cabin at the east end of Isle Royale. This was after many years of very high wolf and low moose density. When I walked into the yard I couldn't believe the trees. There hadn't been many there before, and all of a sudden I could hardly see through the yard, because there was this field of balsam fir just coming up. And that was the release Brian McLaren later documented, which was really striking at the east end. We were surprised later that it also occurred at the west end. And that started me thinking about wolf-induced prey declines and how wolves affect ecosystems.[16]

This scenario was counteracted by what happened next. In 1982 the park's wolf population crashed as a result of an outbreak of canine parvovirus. The moose population shot up to a density ten times higher than on the mainland, illustrating the wolf's strong keystone effect on moose.

In the mid-1990s McLaren and Peterson measured wolf impact across a variety of trophic levels. They cored balsam fir and used dendrochronology to learn what the tree rings had to say about wolf and moose population dynamics. East, west, and middle regions of Isle Royale have different plant communities and growth patterns, best explained by weather and soil variations. The middle of the island burned in 1936 and has few balsam firs. The west end supports very old forests, with sparse but well-distributed balsam firs, overbrowsed by moose. The east end supports the most balsam firs in mixed age classes, and also the highest moose and wolf numbers. Between 1979 and 1994 east- and west-end balsam fir displayed cyclic intervals of ring growth suppression that accompanied elevated moose densities, apparently unrelated to climate or fire. Tree growth was linked to predation, suggesting an indirect wolf effect on balsam fir. When wolves declined the tree rings became narrower, correlated to an increase in moose numbers and an increase in browsing pressure. McLaren and Peterson concluded that top-down control dominated the Isle Royale food chain.[17] However, since this study, bottom-up effects, including climate and moose parasites, have driven balsam fir dynamics as much as wolf predation has. These bottom-up effects may be intensified because Isle Royale is an island and therefore limited in size, and also by lack of all the predator species historically present (e.g., bears in addition to wolves).

Today only fourteen mammal species inhabit Isle Royale. Wolf and moose populations have oscillated widely as a result of disease and predation, leading scientists to conclude that without a full complement of large predators, wolves can't quite control ungulate populations. Long-term studies of these interactions are beginning to yield new insights into ecological relationships. However, Peterson likens scientific findings during any single five-year period of the past fifty years on Isle Royale to the fabled ten blind men describing an elephant:

> Every population goes through a trajectory heavily influenced by unpredictable events, like disease and weather. The longer you study something, the more apt you are to pick up on these meteorite strikes. So if you see these as a one-in-fifty- or one-in-a-hundred-year pattern, and they occur randomly, they are going to show up now and then. I have found that the longer you study something, the less you are likely to be able to package it neatly as an ecological concept, or as a nice, neat ecological interaction that is predictable, repeatable, and generalizable.[18]

Isle Royale continues to yield many lessons about predators and prey, the interaction of bottom-up and top-down effects, and what happens when an ungulate population comes close, as Aldo Leopold put it, to dying "of its own too much."[19]

The year 2008 marked the fiftieth anniversary of the Ecological Study of Wolves on Isle Royale, making it one of the longest-running studies ever done. This event was commemorated in a film by George Desort, *Fortunate Wilderness*, which depicts the astonishing breadth and depth of this endeavor.[20] Over the years other researchers have joined the project, such as population biologist John Vucetich, who has codirected it since 2000. Vucetich's analytical skills have been an asset since the early 1990s, when he became Peterson's assistant. The work of another research team member, conservation biologist Christopher Wilmers, shows that combined effects of disease and climate have led to a shift from top-down to bottom-up moose regulation. Lack of other large predators on this island may contribute to the influence of bottom-up forces. Whether the system will return to top-down control when wolf densities return to higher levels is speculative, but even with higher predation, this may take a while to achieve.[21] According to Peterson, the system has never been top-down or bottom-up but has been a combination of both, with wolf predation the primary influence in the first half of the Isle Royale study and climate dominant in the second half, after wolves succumbed to parvovirus.

As of 2008 the Isle Royale wolf population had recovered to 23 individuals in four packs. The addition of a fourth pack was a bit surprising, since Peterson and Vucetich thought that a decline in wolves was imminent, given a change in moose demographics that included few old animals. The moose population was at 650, the low point in their most recent decline. The story goes on as Peterson and Vucetich continue to measure the ecological forces at work on this island.

The Northern Rockies: Yellowstone and Beyond

The summer before I began my PhD studies, I visited Yellowstone National Park to learn about another grand ecological experiment in which food web effects could be studied, this time through introduction of a top predator. I spent two weeks in the Lamar Valley, an immense glacial valley often called the Serengeti of the American West because of the wealth of wildlife it sustains. The Lamar River meanders through this legendary American landscape, past the dark forms of bison (*Bison bison*) grazing in the sagebrush flats at the valley bottom. A landscape of vast beauty and drama, it holds powerful lessons about the natural world.

I spent some of my time here watching wolves with park naturalist Rick McIntyre. Since the 1995 wolf reintroduction, McIntyre has compiled several thousand pages of field notes on wolf behavior, a valuable scientific and historical chronicle. In addition to his research, he facilitates educational public observation of wolves. This is one of the few places where one can see wolf life unfold in all its drama from a noninvasive vantage point. Several packs range here, easily viewed through spotting scopes from knolls on the valley's northern edge. The context of these wolves' appearances—within a wildland preserve—has enabled people to observe the larger ecological picture of wolf recovery. Over the years, like everything in Yellowstone, these wolves' lives have played out very publicly.

My time in Yellowstone was rather like field biology boot camp. I'd rise at 4:00 a.m., trek up a high knoll on game trails to watch wolves with McIntyre, spend the afternoon examining aspen and willow communities in the valley, and then return at dusk to observe wolves again until the light faded. I pitched my tent next to a stream in a secluded park campground. Most nights I averaged four hours' sleep, too tired to do more than roll over when I heard something big rustling outside my tent in the dark. I observed the Slough pack as they hunted, fed on elk, and reared their young at their rendezvous site, a beautiful high bench across the Lamar River. I spent long hours atop the knoll watching them from a safe point. Well, almost safe. One night our passage down was cut off by a herd of two hundred rutting bison. Someone below radioed to warn us that the bison, which we couldn't see from the top of the steep knoll, were headed in our direction. It took us a long time to weave our way through the herd to safety.

I returned in early autumn and saw the wolf pups, by then almost as big as adults, learn to hunt. Like human children learning about life by playing house, wolves learn to survive as adults by playing hunting games. They made clumsy attempts to take down a bison, which would look at them with disdain and then shoo them away with one good charge. Meanwhile, the adults demonstrated the fine art of evaluating and selecting prey, culling the weak from elk herds. I learned much from these observations, including that if animals drove motorcycles, the wolf would probably drive a Harley. And I learned that Aldo Leopold was right: it *is* all about relationships—although their complexity in a place like Yellowstone is astonishing.

FEW PLACES have been studied as extensively as Yellowstone National Park. Located in northwestern Wyoming and southwestern Montana, created in 1872 by an act of Congress, it covers nearly 3,500 square miles. And few things have been studied as intensively in that park as elk and wolf demographics. The Northern Range, which provides prime elk winter habitat, includes the Lamar Valley. It contains an elk herd that was estimated in the mid-1990s to include 17,000 individuals, an astoundingly large number for the region, but that was down to approximately 7,100 animals by 2009. Between 2000 and 2003 the wolf population here ranged between 50 and 100 wolves per 1,000 square kilometers—among the highest densities in North America—but since 2003 their numbers have fluctuated because of disease. In 2008 the Yellowstone Gray Wolf Restoration Project reported the population at 124 animals, down 27 percent from the 171 wolves recorded in 2007. This is similar to the population decline that occurred between 2004 and 2005, when wolf numbers dropped from 171 to 118 animals.[22]

When Congress created the park, it contained an abundant wolf population. As in the rest of the United States, wolves were steadily eradicated through a poison campaign from the 1910s through the mid-1920s, with no wolves at all reported in Yellowstone by 1928. This had a powerful effect on the ecosystem. In field notes taken in Grand Teton and Yellowstone national parks between 1926 and 1933, Olaus Murie described the range deterioration caused by elk herbivory unchecked by predation. He noted the herbaceous and woody browse species selected by elk at various times of year, observing that in winter and early spring they switched from grass to a diet composed predominantly of aspens. He recorded the overbrowsed condition of the aspens and other woody species and the lack of recruitment. He found this phenomenon so compelling that he visited other elk ranges throughout the West, from Washington to Arizona, looking for broad-scale ecological patterns attributable to lack of predation. Murie's monograph, *The Elk of North America*, would not be published until 1951. In it he reported these effects and commented on the ecological value of predators.[23]

Wildlife management in Yellowstone has gone through four phases, reflecting changes in federal policy, which have strongly affected ecological processes. The first was the aboriginal period of subsistence use of natural resources by native peoples, which ended when the park was established. This was followed by a rocky period of market hunting, attempts at predator extirpation, and supplemental feeding of ungulates, which had irrupted in the 1910s. The third phase began in the 1920s, with culling that continued until 1968 and failed to keep elk numbers at a sustainable level. In 1969 the fourth phase, termed "natural regulation," began. This controversial policy called for allowing natural, or bottom-up, processes (e.g., climate, competition) to control herd size. Predation, or top-down control, was considered "nonessential." The idea was that nature would take its course, and the elk population would reach carrying capacity on its own and then maintain itself at an ecologically optimal level. This was an odd assumption, given that without top predators "nature" was incomplete. The elk explosion that followed caused some of the worst overbrowsing seen anywhere. It was like the Kaibab all over again, except that Yellowstone elk did not crash. One reason there was no crash may be relatively intensive harvesting by humans. Montana initiated a control hunt in 1976, which resulted in the harvest of 8–15 percent of female elk in the northern herd in the years before wolf reintroduction and for a decade beyond.

Because it is considered the crown jewel of the National Park System, or perhaps because this landscape somehow feels larger than life, Yellowstone has always been a place where it seems as if the whole world is watching. A breeding ground for brilliant science, this park also has historically attracted conflict and controversy. Scientists and managers heatedly debated the wisdom of "natural regulation." The National Park Service stalwartly supported this policy, despite lack of supporting science. It would take a keystone species—the wolf—to tip Yellowstone into what ecologist Frederick Wagner calls a fifth management phase, characterized by a combination of top-down and bottom-up regulation via climate change, a drop in elk numbers, and what some consider strong effects by wolves on elk and plant communities.[24] In 1974 wolves received protection under the Endangered Species Act, and in 1995 and 1996 forty-one of them were reintroduced from Canada. The Yellowstone elk population has declined precipitously since then, potentially the result of a combination of factors, including wolf predation and climate variability. This decline is among the factors causing a release of willows (*Salix* spp.), aspens, and cottonwoods, which provide winter food for elk.

Terborgh had already documented the far-reaching implications of carnivore removal in terrestrial ecosystems at Barro Colorado Island and Lago Guri. Peterson's study on Isle Royale identified the wolf as a keystone species. These studies, combined with the natural experiment of wolf reintroduction in Yellowstone, set the stage for multiple independent researchers to investigate the ecological effects of wolves. From an experimental perspective the only thing missing was a "control"—a wolf-free zone for comparison. Park ungulate exclosures provide control areas for aspen and willow studies but are insufficient for research examining complex interactions between predators, prey, and plants. To complicate matters, confounding variables, such as climate change, also became apparent at the time of wolf reintroduction, making it very difficult to tease apart the relative importance of top-down and bottom-up influences on plant growth.

Yellowstone's Missing Aspens, Cottonwoods, and Willows

Ecologists had long been concerned about aspen decline in the park. In the mid-1990s William Romme and colleagues found that few aspens had grown above browse height since the 1930s, with most existing trees having originated in the 1870s and 1880s, before wolf extirpation. He attributed this pattern to a combination of elk abundance, climate variation, fire suppression, and absence of large carnivores.[25] Ecologist Charles Kay suggested that predation and aboriginal hunting may have once limited ungulate numbers. After the release from hunting and predation that occurred when the park was formed, elk numbers rose sharply in the 1920s, excluding other herbivores, such as beavers, that competed for the same resources.[26] Also in the mid-1990s Clifford White did groundbreaking work on aspens in the Canadian Rockies in six national parks. He found a trophic cascade that included humans, wolves, elk, and aspens, with aspens declining in all the parks he studied, and human modification of ecosystems, such as fire suppression, exacerbating the effect of wolf extirpation. All of these studies would inspire Yellowstone research on trophic cascades.[27]

In 1998, two years after the wolf reintroduction, in a now classic *Wild Earth* article, Michael Soulé and Reed Noss commented, in reference to the disappearance of willows in Yellowstone caused by elk overbrowsing, "Beaver, having nothing to eat, abandoned large valleys, and beaver ponds and riparian habitat greatly diminished, impoverishing the local biodiversity. Where wolves have returned, elk herds don't dally as long near streams, and one might hope for the return of the missing beaver ponds."[28] It wouldn't be long before research would support these predictions.

Conservation biologist Joel Berger's 2001 work in Grand Teton National Park provided another seminal study that helped inspire future trophic cascades research in Yellowstone. Berger, an expert on predation risk, and his colleagues looked at a variety of sites where grizzly bears (*Ursus arctos*) and wolves had been eliminated and moose had irrupted. Grizzlies had recolonized some sites. Because habitat structure is one of the best predictors of songbird diversity, he tested how moose herbivory and grizzly predation on moose influence biodiversity from the top down. He found that when grizzly numbers were low, moose numbers were high. Lots of moose meant increased browsing; this decreased available habitat for songbirds, which resulted in lower songbird diversity.[29]

Ecologist William Ripple and his PhD student Eric Larsen conducted the earliest Yellowstone trophic cascades research. They set out to test whether aspen age classes they found were similar to patterns Romme had found and, if so, why.[30] This involved coring hundreds of aspens throughout the Northern Range. They had been discussing the wolf trophic cascade hypothesis as a possible explanation for aspen stand dynamics in the park. Larsen recalled what they found when they counted the tree rings:

> There was aspen regeneration in the Northern Range all the way up to the 1920s, and then it just fell to nothing; essentially, it just fell off a cliff. So I was wondering what happened in the 1920s. I looked at climate and at the fire pattern, and there wasn't any relationship at all. But then I looked at the history of wolves in that park and the disappearance of overstory aspen in the increment core record, and it was a perfect correlation. So I went to Bill and said, "I can't prove this, but this is the only factor that changed to correspond with the decline of aspen." And Bill lit up and said, "Ah, that's really interesting."[31]

Out of that observation sprang a stream of research of enormous consequences. Working in elk winter range inside the park, Ripple and Larsen wanted to confirm the period of 1870 to 1890 as the last major episode of aspen recruitment and determine whether wolf absence was potentially correlated as the result of a truncated trophic structure. As Ripple explained it, "We already had a huge awareness of elk herbivory issues, and then we got the aspen data in, and on top of that we had this third trophic layer, the top predator, and it made perfect sense."[32] Ripple and Larsen's ideas potentially countered "natural regulation," which was based on the concept that elk did not have a negative effect on plant communities without wolves in the system. This policy's proponents argued that aspens were not declining, because they had always been rare in the park.

Ripple and Larsen proceeded to dig deep in their investigation of aspen history, using data from a 1926 beaver study by Edward Warren in which he surveyed existing riparian aspen stands. Warren's data showed that while most aspen had indeed originated in the 1870s and 1880s, there had been recruitment going back to the 1750s.[33] Additionally, Ripple and Larsen found recruitment between the 1880s and 1920, possibly related to elk antipredator behavior. In a subsequent study they intensified their investigation using ungulate pellet counts and wolf radio-collar data. They found that wolves may have influenced elk behavior, causing elk to avoid areas where they had trouble escaping (e.g., ravines, thick forest, places with downed wood). This change in elk behavior suggested the beginnings of a trophic cascade in which aspens were able to grow above browse height as an indirect result of wolf presence. Larsen and Ripple expanded their study, comparing aspen dynamics inside the park and outside, where humans hunt elk, which reduces herbivory. They found a significant difference, with only 6 percent of aspen stands in the park containing trees that originated between 1920 and 1989, versus 84 percent outside the park.[34]

Now a professor at the University of Wisconsin–Stevens Point, Larsen currently studies another keystone species, the southern royal albatross (*Diomedea epomophora*), in New Zealand. However, he has continued to annually monitor the 113 aspen transects he and Ripple established. In 2008 they observed aspen releases in the Blacktail area and in the upper Lamar Valley in the Crystal Bench and upper Slough Creek areas. According to Larsen, in 2003 he never found a sapling more than six and a half feet high, but by 2008 at least 17 percent of their plots had saplings that tall. This may be due to a combination of factors, including change in elk behavior attributed to wolf presence, lower elk numbers that reflect a corresponding contraction in elk winter range, and climate variability.

In a meta-analysis of ungulate density relative to predators, Rolf Peterson found that ungulates commonly achieve high density only when there are fewer than two predator species in a system.[35] Applying these results to Yellowstone, which has multiple predators (wolves, cougars, grizzly bears), Peterson predicts a continuing elk decline until elk numbers reach 4,000 or so. However, their long generations (elk live as long as twenty-six years) create lags in population responses. For example, many of the cows killed in the park today by wolves were born before wolves were present, at the time of maximal elk densities.

Plummeting elk numbers created a change in habitat use. In the 1990s, because there were so many elk, they used all of the Northern Range, including the higher-elevation sites, considered poor peripheral habitat because of deeper snow. As their numbers fell, the elk contracted their presence to the smaller, prime core of their former range. Aspens are releasing outside this core, at the edge of the winter range in the upper Lamar Valley, where there is less herbivory. While this is a density-mediated trophic cascade rather than a fear-based response, the aspen release in the core winter range may be a result of both the elk's fear of wolves and the overall reduction in elk numbers.[36]

Other researchers looked at cascades involving willows and cottonwoods, among them hydrologist Robert Beschta, who visited the Lamar Valley in 1996, just after wolf reintroduction. Beschta recalled finding at that time undercut, barren stream banks. The willows were only ankle high, with very thick bases—so heavily browsed that from a distance it looked as if there were no willows at all. This premier national park's poor riparian condition troubled Beschta for years. He became aware of Ripple and Larsen's work and returned in 2001 to evaluate riparian vegetation. At that time he noticed that the willows had grown taller—to knee height—and in them he saw the first glimmer of a release. But his real breakthrough came when he noticed all the missing cottonwood age classes, which corresponded to the years of missing wolves. Along Lamar Valley streams he found a nearly complete absence of cottonwood recruitment over the previous six decades. He considered factors such as fire history, hydrologic disturbances, and natural stand dynamics; however, none explained the observed long-term absence of new trees. He attributed the cottonwood decline to unimpeded browsing over many years, correlated to wolf absence.[37]

While Beschta was in the Lamar Valley he spent some time afield with Ripple, discussing the beginnings of willow and cottonwood recovery. In 2002 they partnered to produce a remarkable body of research, including more stream work. They found that in the upper Gallatin River basin winter range, heavy browsing after wolf extirpation generated major changes in floodplain functions and channel morphology, adversely affecting habitat for many species, in-

Figure 4.2. Lamar Valley Willow Growth, 1995

Figure 4.3. Lamar Valley Willow Growth, 2003

cluding beavers. Since the wolf's return, improved willow and aspen growth, which helped restore stream habitat, has caused a beaver renaissance, from one Lamar Valley colony in 1996 to ten in 2005.[38]

Top-Down or Bottom-Up?

Scientific debate is alive and well in Yellowstone. By 2010 Ripple and Beschta were two of the many scientists involved in food web research in that park. Not all of them agree that there is a trophic cascade occurring and, if so, what might be driving it. The key question here, as elsewhere, has to do with whether plant releases are indeed caused by top-down effects or whether they are the result of more subtle bottom-up factors, such as climate change and snow depth. Additionally, a spirited debate exists about predation risk mechanisms and whether mortality trumps behavior here, with some researchers questioning the validity of behavioral trophic cascades.

Douglas Smith has been with the Yellowstone Gray Wolf Restoration Project since its inception in November 1994 and has directed it since May 1997. Prior to that he worked on Isle Royale from 1979 to 1992 and was a PhD student of Rolf Peterson. A highly respected and articulate scientist, he is actively involved in park research. He has a gift for cutting through politics and getting to the heart of the matter when it comes to Yellowstone science. Smith refers to the top-down versus bottom-up argument as one of the holiest grails of ecology. Additionally, there is an ongoing debate about whether changes in elk behavior or this species' steep decline since 1995 are triggering cascades. For instance, the presence of fewer elk means less consumption of foods such as aspens and willows, and this is contributing to the vegetation release researchers have been observing.

Both top-down and bottom-up factors influence the size and fitness of any elk herd. In the 2008 annual elk report, park wildlife biologist P. J. White identified top-down factors that caused the drop in elk numbers (predation by wolves and other large carnivores, human harvests of antlerless elk) as well as bottom-up factors (drought effects on maternal health and calf survival). According to Smith, necropsy data show consistent evidence of bottom-up effects on elk health in the park: poor bone marrow, possibly resulting from severe winters and climate change, which may have reduced the quality of forage (grasses). This suggests that wolves are only part of the complex dynamics that affect elk fitness.[39]

Regardless of differences in opinion about the ecological effect of wolves, people agree that this species has had a huge impact. Smith says, "I've started many conversations like this: 'Do you or do you not think that Yellowstone is a different place with wolves?' I've never gotten a no answer. Never. But if you take another step, and say, 'Well, how so?' then the opinions diverge. But nobody will say it's not different." Smith's perspective is that wolves are necessary but not sufficient. Yellowstone, like all ecosystems, is a multicausal system, and wolf presence is not the only thing limiting elk population growth. Smith suggests that to restore natural conditions, you need the dominant North American carnivore. As he explains it, "Here we were trying to do this for decades without wolves. If you want to restore natural conditions in the park, you need to restore, arguably, the most dominant ecological force. I am not arguing against bottom-up, but for the importance of predation."[40]

Some critics think the top-down explanation is a "sexy" hypothesis, well loved by the media, that provides an irresistible explanation for tree recruitment and willow growth in Yellowstone. Others, such as Smith and Peterson, caution that while top-down effects may have prevailed during the first years of wolf reintroduction, the system may be shifting into a bottom-up scenario because of climate change. This does not negate the ecological importance of keystone predators but illustrates the complexity of how their presence interacts with other forces.

Roy Renkin, a veteran vegetation management specialist in Yellowstone, believes willow and aspen growth are primarily influenced by climate. He, ecologist Don Despain, and others are researching bottom-up effects on these species in the park. Renkin cites the 33 percent increase in the length of Yellowstone's growing season since wolf reintroduction, possibly due to global warming, with many more frost-free days. This enables willows to put on more growth per year and produce more carbohydrates for future growth. They can also produce more secondary compounds, which make plant tissues unpalatable to herbivores. By sectioning aspens, Renkin is finding what may be evidence of sustained herbivory in the 1880s, a period characterized by recruitment of this species. According to Smith, Renkin and Despain's findings are valid. However, he emphasizes that top-down and bottom-up hypotheses are not mutually exclusive and suggests that it will take much more study to solve this ecological mystery. As an example of top-down control of aspens, he points out postfire aspen recruitment in 1988 in areas inaccessible to elk, such as scree slopes, while the aspen elsewhere continued to be beaten down by chronic herbivory.

Predation Risk Hypotheses

Predation risk, an imperfectly understood phenomenon, has become a leading component of the Yellowstone scientific debate. To understand better how it works, let's look at several factors associated with predation risk, studied by different researchers, which illustrate some of the facts. In 2003 Ripple and Beschta formulated their *terrain fear factor* idea to explain patterns of vegetation growth on the Northern Range. This provided a conceptual model for assessing the relative predation risk effects associated with elk encountering wolves. They hypothesized that growth of young trees and shrubs would be greatest in riskier areas (sites with a low likelihood of elk detecting wolves and with limited escape routes), and that elk avoid high-risk sites such as those below high terraces or steep cut banks and near gullies.[41]

Scientists are working hard to understand predation risk. As in other systems, trophic cascades mechanisms involve a combination of mortality and behavior. In other words, a release in herbivory could be the product of fewer elk, or the result of elk avoiding certain areas because of fear of predation or other factors, or a combination. While Ripple and Beschta have focused on plant responses to elk behavior, others are using radio-collar data to map elk fear responses and predator-prey interactions. With GPS radio-collar data, which offers a fine-scale look at animal movements, in 2003 landscape ecologist Mark Hebblewhite and colleagues deconstructed predation into three stages: encounter, chase, and takedown. At each stage, various habitat and landscape characteristics potentially influence wolves' hunting success, thereby increasing predation risk. They found that topography largely drives encounters, whereas vegetation type has the greatest effects on risk of death. For example, while odds are high that an elk might be encountered by wolves in grasslands, odds are low that it will be killed there, since it can easily travel through open landscapes. Dense cover in pine forests renders elk more vulnerable to wolf predation because woody deadfall impedes escape.[42]

In 2007 landscape ecologist Matthew Kauffman, Douglas Smith, and colleagues created a predation risk map by looking at where kills occurred. Working with ten years' worth of kill data derived from GPS collars, Kauffman and

Smith obtained results similar to Hebblewhite's. They found that landscape openness enhances wolves' ability to find elk and enables them to more easily sort through a group of elk and identify vulnerable individuals. Elk can easily outrun a wolf under normal conditions. However, obstacles in a landscape (downed wood, debris) slow elk and can improve wolves' hunting success. Additionally, in a system such as Yellowstone, with *so* many wolves, this predator must select habitat not just for hunting success but also to avoid conflicts with other packs. This changes the way wolves organize themselves on the landscape and alters the relationship between predator distribution and predation risk. On a broad scale this means that elk can no longer select habitat to avoid wolves. Indeed, Kauffman and Smith determined that pack territories best predicted elk habitat use. Furthermore, elk increased their winter use of open areas after wolf reintroduction, even though this habitat may be more risky than forested areas in terms of predators detecting the elk.[43]

Smith commented, "On the Yellowstone Northern Range predation risk varies from really high to very, very high. So how can you quantify differences between that? Ten years from now, when our wolf density may be much lower, there may be pockets of low predation risk. But right now wolves go everywhere, because there are so many of them, and predation risk may be high everywhere."[44] He cautioned that predation risk defies easy definition, in part because of the high variation in the landscape that elk share with wolves. Additionally, given the high wolf population, it is unlikely that elk will entirely avoid riparian areas, which contain high-quality browse during critical winter months.

With regard to riparian areas, a further insight into the elements of predation risk came to Smith when he was in the field:

> The point of first contact is never the point where the interaction ends or the wolf kills the elk. There's always a chase. The problem with defining predation risk by the clustering of kill sites around drainages came to me once when I was watching a chase. The elk was clearly outrunning the wolf, and then they hit a steep drainage. The elk had to go downhill, run through some debris, the creek bottom, some rocks, some downed trees, some willow brush, and then run up the other side. By the time it had run up the other side, the wolf had closed the gap. In other words, the wolf

> got through faster than the elk. That particular elk got away. But other times the wolf killed the elk. We had the coordinates for it, and in our database entered it as a riparian area, and labeled it as a drainage with a dead elk. Researchers look at that data and say, "Look at where all the dead elk are." Well, the dead elk are there because of a landscape effect in terms of prey fleeing, not because they were stuck there and couldn't see the oncoming wolves. And that's a very different interpretation. So if I've had an "aha" moment with regard to predation risk, this would be it. Until then I accepted the idea that wolves were encountering elk in riparian areas because they have poor viewshed.[45]

Eighty percent of elk killing in the park occurs at night. This suggests that vision is not a primary sense elk use to detect wolves, and it makes the idea of viewshed (which affects ability to detect predators) less relevant as a predation risk factor.[46] Ungulate expert Valerius Geist cautions that while vision is not elk's primary sense, they use it to detect movement at long ranges, so it can't be ruled out as a component of predation risk response. Additionally, he explains that elk use many strategies to avoid predation. This can involve seeking cover or higher ground, or a combination of responses. Thus what some interpret as contradictory findings may all be part of the complex array of ways that elk respond to predators.[47]

Yellowstone research has inspired work in other places. Ripple and Beschta have replicated their wolf trophic cascades studies in Olympic and Wind Cave national parks in the United States and Jasper National Park in Canada. They consistently found that eliminating top predators caused ungulate irruptions, which led to intense browsing and lack of recruitment. Across a variety of ecosystems they correlated loss of biodiversity and stream erosion to predator extirpation. We will examine some of this fascinating work in subsequent chapters. And in Banff National Park Mark Hebblewhite, Clifford White, and their colleagues did landmark trophic cascades work in which they studied how humans (*Homo sapiens*)—the ultimate keystone species—affect a food web.

The oldest and most studied Canadian park, Banff nestles deep in the northern Rockies amid stunning alpine scenery that includes glaciers, ice fields, and sapphire lakes. It comprises the Bow Valley—a prime, low-elevation elk winter range. The busy Trans-Canada Highway runs through it, and it contains the Banff town site, located in the middle of the park. Wolves recolonized the Bow Valley in 1986, after having been eliminated in the early 1900s, but have avoided the Banff town site because of high human presence. Predictably, this low-wolf area is an elk magnet.

Hebblewhite and White's hypothesis, that human land use has influenced wolf movements and created refugia from predation for elk within the town site, has key implications for resource management. They found wolf predation twice as high outside town, where there are fewer people, with higher sapling density there. This suggests that human-mediated trophic cascades cause substantial negative indirect effects on vegetation, with plant damage increasing significantly in low wolf, high human areas. In willow communities they found decreased beaver presence and decreased songbird diversity in areas heavily browsed by elk.[48]

Serengeti Cascades

For decades Anthony Sinclair has been studying community dynamics in a system to which Yellowstone is often compared—the Serengeti. East Africa's Serengeti-Mara Ecosystem covers 10,000 square miles, three times the area of Yellowstone, and falls in the rain shadow of the Ngorongoro Highlands. This open grassland and savanna system lies within the *Acacia* zone, adjacent to Lake Victoria, mostly in Tanzania. This area is a World Heritage Site because of its legendary migratory ungulate populations, megafauna, and prehistoric sites such as Olduvai Gorge. One of the earth's most diverse systems, the Serengeti contains twenty-eight species of ungulates and ten species of large carnivores.[49] Hardly a "pristine" system, it has been influenced since prehistoric times by multiple human land uses, including fire, timber harvest, and hunting. The rich biodiversity of this grassland, which paleontologists suggest can be used as a reference site for early Pleistocene conditions, and the complexity of the trophic interactions here, which involve slow as well as rapid change, invite long-term ecological studies. Additionally, the low reproductive rates and long lives of the resident megafauna raise concern about their susceptibility to extinction.

A natural experiment involving removal of the majority of carnivores by poaching facilitated Sinclair's study of top-down and bottom-up effects. In the absence of predators, prey populations increased. However, drought and disease outbreaks (rinderpest) caused further prey declines. Reduction in herbivores as a result of these bottom-up effects has led to an increase in trees, providing more cover for lions and improving their hunting success.[50] Each carnivore in this system has preferred prey, typically smaller animals. This selectivity results in smaller ungulates having higher predation rates than larger species and those weighing more than 330 pounds often escaping predation. This causes top-down control of small prey and bottom-up control of larger ungulates.[51]

Sinclair has also studied how migration affects predation success, how fire shapes trophic structure, and the existence of multiple stable states (e.g., grassland and forest) as part of the long, slow ecological changes here. These intricate processes have not been observed in quite the same way in simpler systems, such as Yellowstone. But this may be changing as Yellowstone moves into more of a bottom-up scenario. In the Serengeti, bottom-up processes occur when ungulates' body size is sufficiently large to enable them to escape predation and when herbivores such as wildebeest (*Connochaetes taurin*) undertake large-scale migrations, driven by food availability and climate. Each year during the wet season wildebeest travel from the northwestern corner of the Serengeti, where they graze tall grasses during the dry season, to the southeastern corner of this preserve, where they graze short grasses. A rainfall gradient triggers this movement, which coincides with a green wave of plant growth in this arid region. After the wildebeest have grazed down the grasses into a sward of close-cropped grass, called a *grazing lawn*, they move on, following the moisture to lower areas. This grazing increases the nitrogen in the soil and the protein content of the grazed stems. These effects may have been widespread in North America when migratory bison ranged freely and prior to the extinction of large-bodied species, such as mammoths.[52] Today researchers are beginning to apply the Serengeti's lessons about grassland ecosystems to Yellowstone, to better understand the bottom-up processes affecting elk. And as elk numbers decline, bison may be becoming a more dominant herbivore in that system.

A VARIETY of scientists have identified cascades in terrestrial ecosystems from the tropics to the African savanna, with top-down effects prevailing. However, these effects are not always simple and usually involve interplay of top-down and bottom-up forces, which can operate as in a feedback loop. Predation risk mechanisms represent one of the frontiers for this science. Because of these effects' complexity, some of the debates about how they operate have grown heated. Conservation biologist Michael Soulé suggests that relevant as it may be to parse out these details, we should not lose sight of the forest for the trees, so to speak. Keystone predators drive these effects and should be conserved in as many landscapes as possible for the ecosystem benefits they produce. These benefits include nutrient cycling, healthy soil, robust plant communities, and clean water.[53]

Many of the studies described in this chapter illustrate how removal of keystone predators causes ecosystem simplification and ecological damage. Additionally, it is useful to note that the majority of research on this topic has taken place since 1995 and that the science of terrestrial trophic cascades involving large mammals is still in its early stages. Compelling areas for further research include human shielding (how human presence affects predator activity and how this in turn affects trophic cascades), the effects of climate change on trophic cascades, and trophic cascades outside of protected areas, where multiple human land uses (e.g., livestock grazing, timber harvesting, and hunting) take place. Next we will examine cascades in a type of forest ecosystem so complex that it represents yet another fertile research area with many unanswered questions.

CHAPTER FIVE

The Long View: Old-Growth Rain Forest Food Webs

This chapter is about ecological richness and relatively unexplored terrain filled with possibilities. Considerable trophic cascades research has been done in tropical forests, as in the Amazon, but little has occurred in the Pacific Northwest region of the United States and Canada, which contains ancient temperate rain forests, termed *old growth.* Here we will examine some of the studies from the tropics and then look at Pacific Northwest old growth, including its physical characteristics, our increasing awareness of this forest type's ecological relevance, and potential trophic cascades. We will see how trophic cascades may be occurring at two leading experimental forests: the H. J. Andrews in Oregon and the Wind River in Washington. On the basis of well-documented trophic cascades in other types of ecosystems, we will examine how temperate old-growth forests offer further opportunities to learn about these interactions.

Rain forests occur in the tropics, near the equator, and much farther north, in the Pacific Northwest, as well as in other places, such as Tasmania. Invaluable to our planet's health, rain forests convert large amounts of carbon dioxide into oxygen and provide homes for an astonishing diversity of life. Defined as thick evergreen forests that receive at least eighty inches of rain per year, they cover 7 percent of the earth's land surface and contain as much as 50 percent of known species.[1] They are diminishing rapidly as a result of agricultural conversion and other forms of human land development.

Ecologists characterize old growth, also known as *virgin* or *legacy* forests, as stands more than 200 years old that contain great structural variation. If you take a walk in old growth you will discover trees of all ages, from young to very old; a canopy with multiple levels as you look up into it; standing dead trees (called *snags*); openings made by fallen trees and fires; and abundant dead wood. Openings in the canopy allow light to reach the soil, enabling shrubs and young trees to grow, adding to this system's structural richness. This physical complexity harbors a highly diverse community of plants and animals.[2]

Forests are arranged in layers, ranging from the upper canopy to the soil, each providing distinct habitat. Any mature forest has five horizontal layers, plus one that defines riparian zones. The *canopy*, which has three layers, comprises all foliage, twigs, and fine branches and their flora, fauna, and interstices (the air between them). The *upper canopy* contains the tops of older trees and in coniferous forests the *cone zone*, the area of primary cone production. The *mid-canopy* contains the middle foliage, and the *lower canopy* consists of bottom foliage and sapling tops. Immediately below the canopy lies the *understory*, defined as the area between the ground and the base of the tree foliage, typically filled with shrubs, mosses, and herbaceous species such as wildflowers. The soil forms the base layer, made up of inorganic material and plant roots and inhabited by insects, small mammals, and mycorrhizal fungi, which have symbiotic relationships with tree roots. Finally, the *hyporheic zone* makes up the below-ground hydrologic system fed by stream water. It provides a spongelike interface between terrestrial and aquatic regions where nutrient exchange takes place.[3]

Ecologists Gary Polis and Donald Strong refer to trophic cascades in systems with high biodiversity as "trophic trickles" because food web complexity across a spectrum of multiple herbivores, carnivores, and detritivores (species that obtain nutrients by consuming decomposing matter) creates many pathways for energy movement. We saw in chapter 3 how coral reefs provide examples of this. Old-growth rain forests, whose diverse structure encompasses whole watersheds, provide another example. Accordingly, we'll look at food webs at various levels in these forests, from the top of the canopy to inside the soil.

The Empty Forest

In the 1990s conservation biologists Kent Redford and John Terborgh identified a phenomenon called the *empty forest* whereby apparently rich tropical forests lacking keystone and other animal species undergo changes in plant communities. These effects move through a food web, influencing everything from tree growth to soil function to wildlife habitat, and can be as damaging as deforestation. Plant community changes occur anywhere large mammal populations have been overharvested by humans or reduced or eliminated by climate variability. These effects can have long-term influence; indeed, many forests reflect events that happened thousands of years ago.

If you walk through a Costa Rican rain forest you will see hundreds of large fruits lying on the ground rotting, their seeds failing to germinate. Scientists puzzled over this mystery for years. The forest palm *Scheelea rostrata* grows in lowland rain forests from Brazil to Costa Rica and produces yellow, egg-sized fruits with thick-hulled seeds. Because many things in nature have evolved to ensure survival of species, it made no sense that this tree had developed seeds that could not germinate. In a seminal 1982 *Science* article, "Neotropical Anachronisms: The Fruits the Gomphotheres Ate," evolutionary ecologists Daniel Janzen and Paul Martin looked at this puzzle through the lens of evolutionary time. They began by considering that all species in an ecosystem coevolve, developing traits that serve specialized functions.

The food web in this landscape once contained gomphotheres, mastodon-like Pleistocene herbivores, which ate the fruits of the large, relatively long-lived forest palms. Gomphotheres ingested these fruits and moved on, probably traveling far each day. During the slow digestion process they excreted the seeds far from the trees where the fruits originated. The gomphotheres' movement patterns distributed the seeds broadly, and enzymes to which the seeds were exposed during digestion enabled them to sprout. Gomphotheres became extinct about 10,000 years ago, possibly as a result of overhunting by humans. In their absence there has been no other large mammal capable of ingesting and dispersing forest palm seeds. Janzen and Martin's megafaunal extinction hypothesis, in which they postulated that these seeds coevolved tightly linked to gomphothere feeding habits, helped solve this mystery. They further suggested that many similar evolutionary trophic relationships may have been lost because of extinction, possibly creating forests that today are different from those that would have existed had these species not become extinct. With megafauna to disperse their seeds, these forests originally may have been more dense, with a broader distribution.

Janzen and Martin's paper invited scientists to look at trophic relationships on longer time horizons. It had a profound effect on marine ecologist James Estes, inspiring his research on the plant defense compounds kelp forests produce in the Southern Hemisphere, where there are no sea otters (*Enhydra lutris*) to keep herbivores in check.[4] When I read their paper I realized that the patterns and processes I had been observing in nature had far deeper origins than I had imagined. And had I read it before I traveled to the Amazon, I would have experienced that system differently.

The Amazon Basin encompasses 3 million square miles and nine countries and has the highest biodiversity of any region in the world. I visited the Peruvian portion in the mid-1980s, traveling downriver from Iquitos in a dugout canoe and going deep into the rain forest. Since then urban growth has caused deforestation via cutting and burning, but I had the good fortune to see this area before it was logged. My first impression was of unparalleled sensory richness and the sharp contrast of the benign and the deadly—pink river dolphins (*Inia geoffrensis*) swimming boisterously next to our canoe, and cayman alligators (*Caiman crocodilus*) gliding silently toward us, their olive bodies and rugose snouts camouflaged against the muddy water.

I spent halcyon days at a primitive field camp, eating exotic fruit and feeling euphoric from breathing the moist, superoxygenated rain forest air. Not long after I arrived a man from the Yagua Tribe, who stood four and a half feet tall, carried a blowgun, and wore little more than a loincloth, led me into the forest. My barefoot guide started on a narrow, overgrown trail, gracefully pushing aside tangled vegetation and palm fronds as he penetrated the forest. We were soon enveloped in verdure, and the dim sunlight that filtered down from the canopy far above created a primeval atmosphere. Ancient trees soared skyward, taller than the nave of any cathedral, and held wonders beneath their interlaced canopy: enormous spiderwebs; boa constrictors (*Epicrates cenchria*) draped on low-lying tree limbs, their rust-and-black bodies barely visible in the half-light; and toucans (*Ramphastos toco*) streaking by, their yellow bills and breast feathers bright against the soft green foliage. The primal cries of howler monkeys (*Alouatta* spp.) provided a rousing accompaniment for our walk.

The hardwood trees and forest plants looked unlike any in the Northern Hemisphere. The massive buttress roots of mahogany (*Swietenia macrophylla*), which Peruvians call *caoba*, supported thick, towering boles, and Brazil nut trees (*Bertholletia excelsa*), the largest of the Amazon rain forest species, rose unbranched for half of their 150 feet of height. Vines called lianas coiled around tree trunks and hung from high branches, creating pathways for wildlife. Gardens of epiphytes (plants that do not need to grow on the ground) sprouted from stout limbs, some of them orchids, others bromeliads, many in bloom, their heady, sweet scent permeating the air. They had evolved mechanisms to make the most of this environment, collecting moisture and soil in their exposed, tangled root masses and cupped leaves.

I was not an ecologist back then, so I missed much, especially the signs of damage in this forest, even amid all the lush beauty. Back then I believed forests could be destroyed only by chain saws and the slash-and-burn swidden agriculture prevalent in South America. I didn't realize that large predators, such as the jaguar (*Panthera onca*) and puma (*Puma concolor*), were most likely gone from this place and that their departure had already caused great changes. In the years after my Amazon sojourn, through the writings of Janzen, Martin, and Redford, I became aware that damage to tropical rain forests can occur from within, less visibly and more insidiously than damage wrought by logging.

Redford investigated the truncated trophic cascades created by human harvest of animals in Amazon rain forests. These stands look deceptively dense and jungle-like, but when one looks closer it becomes apparent that they are empty.

Animal removal, termed *defaunation*, can be direct, by hunting or habitat destruction, or indirect, by harvesting of the fruits wildlife need for survival. Both occur regularly throughout Latin America. Redford estimated that in Brazil human subsistence hunters remove 14 million individual mammals per year. Current hunting is much higher than it has been historically because of the market for bush meat (wild meat). Hunters harvest monkeys and other jungle animals to meet the high food demand in logging and mining camps, driving to near extinction the large predators and herbivores that perform the important role of seed dispersal. Redford described the resulting forests as composed of "living dead" tree species, which will die without replacement. The harvested animals' absence truncates food chains and also reduces seed predation, herbivory, pollination, and predation on other animals. In Amazonia a forest open to hunting contains carpets of seedlings so thick that intense competition prevents all but a few from surviving, as well as piles of uneaten, rotting foods and ungerminated seeds. In contrast, a forest without hunting contains recruiting trees of large-fruited species, complex stand structure, and intact food webs.[5]

In southeastern Peru's Río Manú floodplain, John Terborgh and colleagues assessed how hunting alters recruitment of trees into tropical forest canopies. He compared Boca Manú, a site where large vertebrates had been overhunted, with Cocha Cashu, a reserve that contained intact fauna. Boca Manú species that had been greatly reduced included arboreal seed dispersers, such as spider monkeys (*Ateles* sp.), howler monkeys, and white-faced capuchins (*Cebus capucinus*), and terrestrial seed predators, such as collared and white-lipped peccaries (*Tayassu tajacu* and *T. pecari*). Terborgh measured sapling recruitment, seed dispersal, and the prominence of large-seeded species among those that showed impaired recruitment. He found few tree species populations that were stable, with 75 percent decreasing. Over time removal of animals is likely to diminish biodiversity, but this trend may be reversible if hunting is prohibited. Terborgh's research demonstrated that vegetation change in response to altered animal communities represents a serious threat to biodiversity, particularly because these effects may be subtle at first and difficult to detect without systematic surveys of plant and animal communities, until an ecosystem has passed a tipping point into another phase state.[6]

Pacific Northwest Forests: From Sustained-Yield to Old-Growth Conservation

A temperate old-growth rain forest is among the earth's most species-rich systems. However, this diversity is not uniform. Some taxa are rich, and some are not. For example, temperate old growth has few tree species but lots of fungal species, the reverse of tropical forests. High influx of nutrients and high standing crops of trees support this diversity, as do old-growth physical characteristics, as we shall see.

Just as a coral reef provides the structure and home for myriad species, in the Pacific Northwest the Douglas-fir (*Pseudotsuga menziesii*) serves a similar function, so much so that eminent forest ecologist Jerry Franklin and others refer to it as a keystone species.[7] While this designation does not strictly fit Robert Paine's definition of a keystone (i.e., a carnivore), this tree undoubtedly defines Pacific Northwest rain forests. Not a true fir, but rather the largest member of the pine family, the Douglas-fir has a colorful taxonomic history. Its genus name, *Pseudotsuga*, meaning "false hemlock," refers to the fact that early botanists misclassified it. This fast-growing generalist species ranges from southern Mexico to central British Columbia, growing from sea level to 9,000 feet. One of the most fire-hardy conifers due to its thick, corky bark, it has flat needles that are irregular in length and large cones with distinct scales that end in three-pointed bracts. Its branches grow in whorls tipped by sharp-pointed buds. This generally shade-tolerant species can regenerate in the understory of other trees, including pines. A formidable giant, it can grow to 300 feet and live for more than 1,000 years.

A Douglas-fir's high, multilayered canopy, thick bole, and enormous biomass, whether it is living or a snag, provides home for many organisms, such as the Douglas squirrel (*Tamiasciurus douglasii*), which harvests great quantities of cones, and the northern subspecies of spotted owl (*Strix occidentalis caurina*), which nests in broken-topped trees and in cavities abandoned by other birds. Remove the "keystone" Douglas-fir and the system ceases to exist as we know Pacific Northwest rain forests, although in time you might still end up with a coniferous forest made up of other species, with quite a bit of old-growth structure. Douglas-fir also yields more timber in the United States than any other tree, which adds another dimension to our human relationship with it. If you drive through Washington and Oregon you will see rows of young Douglas-firs growing vigorously in privately owned cutover lands west of the Cascade Divide, waiting to be harvested.

In any discussion of food webs in temperate forests, it helps to begin with the concept of old growth and consider how it arose and what it has come to mean to science and conservation. For eighty years forest management in the United States was based on sustained yield, a silviculture style developed in Europe in the mid-1800s. Brought to this country by Gifford Pinchot, the first chief of the US Forest Service (USFS), it called for growing trees as crops and harvesting them sustainably. This meant establishing a logging schedule based on the time it took for new trees to grow and replace the harvested trees. In the Pacific Northwest, the first half of the twentieth century was one of stewardship by the Forest Service because private interests did not want federal timber in the marketplace as they cut their own holdings. At this time old growth was the leading forest type targeted for logging because of its profitability. After World War II the federal timber era took off and old-growth harvest escalated, with an emphasis on sustained yield, despite environmental policy. But by the 1970s growing ecological awareness and unease about the conservation status of our natural resources led us to reevaluate the sustained-yield model and the principles we were using to manage forests.[8]

Early ecologist Frederick Clements' concept of succession and the climax community provided the theoretical foundation for sustained yield. His ideas created the dominant model for how ecosystems functioned until Charles Elton came along with an alternative perspective. In the 1920s Clements proposed that without human intervention, succession, or natural progression from one ecological stage to another, drove community dynamics. After a disturbance such as fire, flooding, or glaciation, all communities strove to the end point in their development, also referred to as the climax or equilibrium state. Climax communities, in which individuals are replaced by others of the same kind, are relatively stable. They feature a wide diversity of species and complex food webs. Clements suggested that succession to the climax was a predictable outcome and that climate (i.e., the recession of the glaciers) had an important role in this dynamic.[9] Additionally, he depicted succession as a universal process that creates patterns in the composition of any type of forest. Clements' ideas are quite valid today, but they explain only part of what drives community ecology. Today we know that the development of a forest from one stage through the next, via succession, provides a guiding force that shapes this system alongside other factors, such as trophic cascades.

Since understanding succession is foundational to understanding forests, let's follow a typical forest along its successional development. Each forest type, such as eastern hardwood or boreal conifer, has its own succession pathway, marked by certain species giving way to others. For example, after a stand-replacing event such as a wildfire, a pioneer forest of young Douglas-fir saplings comes up. This is referred to as an *early seral* community—an early stage in ecological succession as a community advances to the climax stage. It is also referred to as the *early successional* phase in the maturity of the stand, a time of heightened biodiversity due to resource availability (sunlight). The trees grow rapidly for several decades until the canopy begins to close, preventing light from reaching saplings on the forest floor, and biodiversity drops. Silviculturists refer to this as the *stem exclusion* phase. This enables species that are more shade tolerant, such as western red cedar (*Thuja plicata*) and western hemlock (*Tsuga heterophylla*), to sprout and thrive. Competition is a key aspect of this process, with certain organisms giving way to others on the basis of how much sunshine and moisture are available. Centuries pass, the long-lived Douglas-firs age and begin to die, and forest composition progressively shifts to other species, which eventually dominate the forest. As the forest matures further, biodiversity increases again because of increased diversity in stand structure. In the climax stage the forest primarily consists of western hemlock. The time required for succession from a young Douglas-fir stand to a climax hemlock forest can be as long as 1,000 years. Many forests never reach this stage because of disturbances that interrupt this process.

Clements' ideas about succession became popular during the 1930s, when the Forest Service was struggling to address economic problems caused by the Great Depression. We had been through other major upheavals, such as World

War I and the 1918–1919 influenza pandemic. Living in what appeared to be a deeply unstable, troubled world, people looked to science to help provide a path to stability.[10] Clements' ideas would influence timber harvest by supporting the notion that forests grow back after disturbance, allowing managers to reharvest this resource in an indefinitely renewable manner. But even back in the 1920s and 1930s alternative views existed.

British ecologist Charles Elton, who created the food pyramid and other seminal ecological concepts still used today, believed ecosystems were governed by complex interactions that worked alongside succession to create additional dynamics, such as the top-down effects of predators. In contrast to Clements' predictable succession patterns, Elton asserted that ecosystems were characterized by complexity, unexpected consequences, and the unimaginable—what today we refer to as *environmental stochasticity*. His ideas influenced Aldo Leopold, who as a 1940s resource manager pointed out the importance of acknowledging nature's uncertainty.[11]

In the 1970s the Forest Service continued to apply sustained-yield principles and considered ancient forests to be "decadent"—past their prime and far less productive than younger forests, which grew more vigorously. Nobody had yet defined the concept of old growth, much less recognized its ecological value. Jerry Franklin, then a young scientist on fire with new ideas about how forests worked, became the first to do so. His early work unfolded at the H. J. Andrews Experimental Forest, one of eighty-one forests set aside by the Forest Service for research purposes.[12] The science produced here would provide the wellspring of what became known as the "new forestry"—a more holistic approach to forest management. This progressive, ecosystem-based way of looking at forests would enable government agencies to incorporate growing knowledge about the effects of harvesting our natural resources and of natural disturbances on native ecosystems, too, such as the Mount St. Helens eruption. It led to the policy of ecosystem management, as part of the Forest Service's New Perspectives program in the early 1990s, and today may be informing the way we manage keystone species on public lands. We will explore ecosystem management further and its application within a trophic cascades framework in chapter 8.

Long-Term Ecological Research: The H. J. Andrews Experimental Forest

The H. J. Andrews Experimental Forest (HJA) is part of a network of living laboratories established by the Forest Service. Founded in 1948 and located fifty miles east of Eugene, the HJA encompasses 16,000 acres deep in the Oregon Cascades, in the Blue River drainage of Willamette National Forest. It ranges from 1,380 to 5,350 feet in elevation. Douglas-fir and western hemlock dominate its lower regions, along with western red cedar, with Pacific silver fir (*Abies amabilis*) growing at higher elevations. Lookout Creek, a tributary of the Blue and McKenzie rivers, runs through this watershed, its clear waters providing habitat for cutthroat trout (*Oncorhynchus clarkii*) and the coastal giant salamander (*Dicamptodon tenebrosus*). The Forest Service's Pacific Northwest Research Station, Oregon State University, and Willamette National Forest administer this forest jointly, with a history that includes Jerry Franklin, Frederick Swanson, and most recently Barbara Bond as principal investigators. Natural history writer and lepidopterist Robert Michael Pyle referred to the HJA as a place where "when a tree falls in the forest, a lot of people hear it—and then take a close look at what happens next."[13] Part of the National Science Foundation's Long-Term Ecological Research (LTER) program since 1980 and a UNESCO Biosphere Reserve, the HJA has led forest ecosystem research for decades.

The LTER network comprises twenty-six sites, mostly in North America, with one in Antarctica and another in the western Pacific on non-US territory, that exemplify a wide range of ecosystem types, from kelp forests to urban landscapes. At each site scientists focus on five core research areas: primary production, distribution of populations selected to represent trophic structure, accumulation of organic matter, nutrient movement, and disturbance. At the HJA the overarching question has to do with how land use, natural disturbances, and climate variability affect key ecosystem properties. Addressing this requires research that potentially spans several generations of human lives, that is, the time it takes for a log to decompose or a young forest to develop into an old-growth stand.

Ecologist Mark Harmon's log decomposition study, which is taking place on a monumental scale, exemplifies this long-term approach. Since 1985 he has been studying how trees rot in this forest, in plots that he intends to be surveyed beyond the next century by generations of future ecologists. Operating under the assumption that each log can be considered an ecosystem on its own because of the amount of life it sustains, Harmon suggests that these relationships continue long after its death, until it has become transformed into other states, such as soil carbon, organic matter, or gas—in other words, until it has decomposed beyond recognition as a log. To quantify this he is measuring the amount of carbon dioxide a decomposing log gives off and the nutrients, insects, and fungi it harbors, as well as its hydrologic function as a water reservoir on the forest floor.[14] HJA studies such as this aim to increase understanding of the ecological function of old growth. As Pyle put it, "The long view requires faith in the future—even if you won't be there to see it for yourself . . . looking to the future is a way of hoping there will still be something to see when we get there. Maybe it's the only way to make sure of it."[15]

Much of what goes on in old growth takes place in the hidden forest: the soil. For example, *mycorrhizae* are fungi that form a symbiotic relationship with the roots of a plant. Trees provide them with nutrients in the form of carbohydrates from photosynthesis. In exchange these fungi act as extensions of the tree's root system, aiding its uptake of water, nitrogen, and phosphorus. Food web processes such as these are vital to forest health.[16]

In the 1970s James Trappe, a fungus expert, and Chris Maser, a mammalogist, discovered a mutual relationship between the endangered western red-backed vole (*Myodes californicus*), truffles (*Tuber* spp.), and Douglas-firs. Truffles are underground versions of mushrooms, the "fruit" of the fungal mat created by mycorrhizae. Roughly round and brown, they resemble small potatoes and range from one-half to three inches in diameter. Because they grow entirely belowground, they can disperse their spores only by being eaten, in a process called *mycophagy*. The red-backed vole subsists almost entirely on truffles, which also provide an important food source for other animals, such as the northern flying squirrel (*Glaucomys sabrinus*) and even mule deer (*Odocoileus hemionus*) and elk (*Cervus elaphus*). These animals eat the truffles, in this manner dispersing spores essential to the growth of Douglas-firs and other conifers.

And when a Douglas-fir dies and falls to the ground, it rots, creating a nutritious medium for truffles to grow in, completing this cycle.[17]

Disturbance, which influences every aspect of a forest, including food webs, provides another leading research topic at the HJA. Frederick Swanson has spent a lifetime studying disturbances ranging from the 1980 Mount St. Helens volcanic eruption to landslides in forest streams. Tall and lanky, he has a full beard and the sharp eyes of one who notices and carefully considers everything. When he speaks about disturbance his face grows animated and he starts gesturing with his hands. He began his HJA work in the 1970s with a new PhD in geology, doing a postdoctoral study with Franklin on how disturbance affects landform development. From the start he saw the HJA as a dynamic ecosystem where mountains, trees, and streams were always moving, sometimes slowly, sometimes in catastrophic bursts, as in the sudden flow of muddy debris after a long rainstorm. Debris flows can occur as a result of exceptionally heavy rainfall, which causes streams to rise. Generally debris flows start as saturated soil breaks loose from hillslopes and cascades downslope, into steep streams, and then down the river channel. The resulting torrent breaches natural dams, moving dirt, rocks, and fallen trees down hillslopes and streams, scouring the channel. This can widen a stream, causing large changes in the landscape and plant communities. In 1996 heavy rains and snowmelt sent giant logs and boulders hurtling down Lookout Creek, providing a tremendous opportunity for Swanson to observe these geomorphic dynamics up close. Over the years he has broadened this work to include volcanic activity in far-flung places, such as Chile, and other disturbance types, such as fire and windstorms, and has also become engaged in policy work related to forest management.[18]

The Guru of Old Growth

The 1970s and 1980s were exciting times at the HJA, with Franklin assembling an interdisciplinary team to conduct cutting-edge research on forest science topics that ranged from disturbance to nutrient flow. Currently holding an endowed chair, and a professor of ecosystem analysis at the University of

Washington, he is referred to as the "guru of old growth" because of how his revolutionary science has changed forest management over the years. On a warm, humid midsummer day in 2009 I took a walk in the woods with him in Olympic National Park, along the South Fork of the Hoh River. We followed a trail through one of his favorite groves of centuries-old spruce and fir, discussing his early work at the HJA and the intricacies of food webs in this forest. Winter wrens' lilting melodies echoed through the understory created by the giant trees, providing a backdrop for our conversation. A veteran forest scientist, Franklin knew these woods well and moved easily through them as he showed me the complexities of old growth. And as we explored this stand we became engaged in a lively discussion about the likelihood and unlikelihood of trophic cascades in this system.

Franklin has always been drawn to uncharted terrain. In the early 1970s, supported by the National Science Foundation, he turned his full attention to old growth. Until then no formal definition of this forest type had existed, yet emerging forestry practices required that managers set some old growth aside.[19] To that end he convened the HJA researchers and led them in creating a monograph titled *Ecological Characteristics of Old-Growth Douglas-Fir Forests*. Published in 1981, it would provide the basis for policy that transformed traditional forestry. Franklin and his colleagues proposed that old-growth forests differ significantly from younger ones in species composition and in the ways they function, that is, rate and paths of energy flow, nutrient and water cycling, and physical structure.

Four fundamental structural components create an old-growth forest's complexity: large live trees, many snags, and logs on the ground and in streams. In their report Franklin and his colleagues referred to old growth as far more than a collection of large, decadent trees. They saw dead, decaying organic matter as an important resource, to be maintained for the ecosystem benefits derived from it, such as enhanced nutrient cycling and fish habitat. Additionally and most radically, this report proposed an ecosystem approach to resource management, one in which managers recognized the interlinked nature of forest components, which included organisms and their functions. In this scenario, successfully managing old growth meant avoiding salvage logging after a wildfire, to allow snags to remain standing and provide a home for wildlife. The HJA team recommended that even within a multiple-use framework, managers acknowledge the nontimber value of individual trees. They also suggested maintaining trophic interactions among species at all structural levels.[20]

In the decade that followed, Franklin and a core of researchers, including Frederick Swanson, stream ecologist Stanley Gregory, and others, worked hard to define patterns and processes in old-growth forest. Gregory began his work here in the 1970s, first studying primary productivity (energy flow driven by sunlight, moisture, and photosynthesis) in streams. Later he began investigating the role of logs in streams and the interface between riparian areas and land. He saw streams as three-dimensional interaction zones bounded by a river's floodplain and the forest canopy, with disturbance creating a variety of habitats, determining the abundance and quality of nutrients, and influencing trophic relationships.[21] Also, he led production of a stream and riparian management guide for the Willamette National Forest plan of 1990, which became a foundation piece for later conservation strategies, such as the Northwest Forest Plan. This and other benchmark research helped people begin to see ecosystems in new ways, to the extent that in the 1990s the HJA would be the epicenter of one of the most intense environmental controversies in US history.

The Northern Spotted Owl and the Northwest Forest Plan

Research on the northern spotted owl (*Strix occidentalis caurina*) began in the 1970s, led by Eric Forsman, a master's student, working in the HJA and in neighboring forests. Endemic to the Pacific Northwest, this subspecies of the spotted owl has white spots on its head, neck, and back; brown bars on its chest and abdomen; and big, dark eyes set into a prominent facial disk. It nests in the tops of old trees and snags, mates for life, tolerates disturbance poorly, and feeds primarily on wood rats (*Neotoma* spp.) and flying squirrels.[22]

In the 1980s Forsman was the first to determine this owl's habitat needs: old-growth forest, a community type that by then had been reduced severely in the Pacific Northwest.[23] Habitat fragmentation and reduction caused the northern spotted owl to be placed on the list of species to be considered for government protection. Conflicts escalated on federal lands over protecting its habitat, which meant reducing timber harvest. In 1989, in response to growing concerns, the federal government asked wildlife ecologist Jack Ward Thomas, who later became chief of the Forest Service, to create an independent scientific plan for the owl's recovery. For political reasons, which included impacts on timber harvest, the plan was never formally approved; however, in 1990 the federal government listed the owl as threatened under the Endangered Species Act (ESA). In 1991 Congress chartered the Scientific Panel on Late-Successional Forest Ecosystems to develop multiple alternatives for this subspecies' conservation. The panel's four principal scientists, who came to be called the "Gang of Four," were Thomas, Franklin, forest policy expert K. Norman Johnson of Oregon State University, and John Gordon of Yale University.[24] To meet ESA criteria for population viability of wildlife dependent on old growth, in 1993 the Forest Service drafted the Northwest Forest Plan (NWFP). This statute took an ecosystem approach to forest management using the best available science.[25]

According to Thomas, in the five years between 1989 and 1994, as a result of changing environmental policies to protect old growth, management objectives on public lands within the spotted owl's range shifted dramatically from sustained yield of timber to maintenance of biodiversity, with an emphasis on endangered species.[26] This resulted in wholesale reduction of logging of old-growth forests. Much of the remaining old growth was designated for protection in *late successional reserves* on 24 million acres in the Pacific Northwest under the Northwest Forest Plan. Almost overnight timber yield plummeted by 80 percent, although in practice it may have been more like 90 percent.[27] Johnson has referred to this as a no less catastrophic regime shift than the Mount St. Helens eruption. It involved an abrupt transition from Pinchovian forestry to the holism of the "new forestry," with this owl the flagship species for old growth. This precedent-setting conservation story provides a compelling example of the relevance of taking an ecosystem approach. And it opened the door for future science-based resource management, such as policies involving keystone species (e.g., gray wolf reintroduction), which we will examine in chapter 8.

Another Empty Forest?

We have seen how the absence of large mammals in tropical rain forests left plant communities simplified. Now let's examine how these effects may be functioning in the Pacific Northwest from the perspective of wolf (*Canis lupus*) conservation. Humans had effectively removed wolves from Oregon by the mid-1940s because of fear of how they would affect livestock and elk and deer populations. However, by the late 1990s wolves had begun returning from Idaho. In 1999 a radio-collared female arrived in the eastern part of the state, and in 2008 the Oregon Department of Fish and Wildlife confirmed wolves denning there.[28] As of 2009 wolves were not known to exist at the HJA, in central-western Oregon; however, given their peripatetic nature, it may be only a matter of time before they recolonize this general area as well.

Like many western states, Oregon has a long-standing reputation for abundant wild game. While what we know about historical elk numbers is somewhat speculative, it appears that they were high in the early 1800s but had declined by the end of the 1800s as a result of overhunting by humans. As a result, in 1899 the state legislature responded by making it illegal to sell wild animal meat, banning elk hunting from 1909 to 1932. In the 1930s this species' population began to increase, very likely because of hunting restrictions and predator extirpation as well as logging, which improved habitat by creating forest openings filled with browsable sprouts.

The HJA harbors a wealth of invertebrates but low numbers of large mammals. Oregon has two native subspecies of elk: Rocky Mountain elk (*Cervus elaphus*), which inhabits the eastern part of the state, and Roosevelt elk (*C. elaphus roosevelti*), which lives in rain forests. Current elk presence in the core of HJA old growth is negligible because this species tends to prefer open habitat, such as logged areas, and the edge, or ecotone, between forest and grassland. The elk population in what is now the HJA may have been higher when that area was a younger forest, but as in other areas, we have insufficient archaeological information, such as remains of human encampments and fire circles containing elk bones, to be certain. Other large mammals currently here include mule deer (*Odocoileus hemionus*), coyotes (*Canis latrans*), cougars (*Puma concolor*), and black bears (*Ursus americanus*). Cougars and bears exist in low densities, but even with these carnivores present, the resulting food web has been truncated in an area that once held a wolf population. Historical references, such as the predator surveys conducted at the turn of the century by US Bureau of Biological Survey biologist Vernon Bailey and others, provide ample evidence of wolf presence here. And these historical reports make one wonder what this place might have been like when it had an intact predator guild.[29]

I spent some time on a cold, cloudy spring day walking through the old-growth forest near HJA headquarters with Swanson and hydrologist Julia Jones, discussing this question. Both know this forest with an intimacy born of years of research. Jones studies stream flow response to forest harvest and landscape-scale disturbance patterns. Petite, with wavy brown hair and an infectious smile, she has made major contributions to the research in this forest. And so I was well accompanied as we set out to look for trophic cascades. In the absence of any formal research on the food web effects of large mammals here, we discussed the possibilities. While our conversation that day amounted to little more than conjecture, the idea of missing fauna and the changes this might have caused in plant communities intrigued us.

Swanson led us into the forest on a narrow, snow-packed trail. He clambered over the massive, mossy trunks of fallen Douglas-firs in our path, moving with the speed and agility of one who knows this landscape well. As we walked I examined the vegetation, which did not appear to be suffering from too much herbivory, because elk and deer numbers were low. But taking a longer view, Swanson suggested that the apparent absence of a strong trophic cascade at the HJA may be a reflection of today's forested landscape, which includes the properties of an old-growth forest, such as great structural and species diversity, as well as its properties on a longer time scale (millennia), which have to do with this system not having a tipping point between phase states, the way a kelp forest or a forest-grassland system in Yellowstone National Park does. However, in considering old-growth dynamics, he suggested that there may be periods when a cascade could have occurred here—for example, in the immediate aftermath of an extensive, severe fire.

Elk follow fires because a high density of tender, nutrient-packed shrub and tree sprouts come up during the first few years after a fire. This may explain why HJA forest ecologists, who have found a connection between tree regeneration and site disturbance histories (e.g., wildfire, defoliation of Douglas-firs by insects), have been observing a historical lag (approximately forty years) in tree establishment following disturbance, on the basis of tree ring data that goes back to the 1500s, with some data going back to the 1400s and 1300s. This lag could possibly have left some time for expansion of elk herds in the early seral periods of the early 1500s. Other factors that may influence a trophic cascade at the HJA include differential browsing pressures on plant species, based on their taste appeal to ungulates, and the time necessary for an ungulate population, which may normally be low in old growth, to increase after a disturbance in order to capitalize on enhanced resource availability.

As in other places, it's likely that removal of large carnivores may have left broad, landscape-scale marks on this forest via trophic cascades. However, our understanding of how this may have worked in this ecosystem remains uncertain. Swanson emphasizes the need for researchers to develop a general conceptual model of these interactions—one that incorporates potential effects of the logging and broadcast burning prevalent in the early 1900s, as well as other forms of human-caused disturbance, including aboriginal use of resources. Ideally this model of vegetation change and trophic cascades would, like most of the work being done at the HJA, employ a long-term, multicentury approach. Accordingly, it would consider natural and human influences on distribution of early seral habitat. This forest type has waxed and waned in distribution over the past five hundred years for which we have spatially explicit records.

Additionally, Franklin and Swanson suggest that it's likely that a Douglas-fir–western hemlock forest may lack a keystone predator, as in simpler systems with shorter successional pathways. While wolves may have historically exerted a top-down influence in some localized areas on a limited time scale, with disturbance a factor (i.e., disturbance creating forest openings, which draw ungulates, which then draw predators) and plant community patterns reflecting this, overall it is unlikely that a top predator would have the same sort of controlling influence on the whole system as it does in the northern Rockies in aspen communities. Part of this may be because of the nature of Douglas-fir–western hemlock successional dynamics. At any rate, it would be important to science to investigate these dynamics.[30]

Old-Growth Reflections: The Spring Creek Project

In addition to world-class science, the HJA also fosters innovative cross-disciplinary work. Swanson's view of this place as more than the sum of its parts has helped create a fruitful collaboration with the Spring Creek Project. Affiliated with Oregon State University and directed by philosopher Kathleen Dean Moore and poet Charles Goodrich, this program brings together environmental sciences, philosophy, and writing to find new ways to envision our relationship with nature. This approach is as relevant today as it was in the 1930s, when Aldo Leopold (a writer and scientist) and Olaus Murie (a painter and scientist) used the arts to inform their scientific work. At the HJA, the LTER program and Spring Creek's Long-Term Ecological Reflections program create regular opportunities for writers and scientists to work interactively in the forest and tap into each other's wisdom, yielding a rich body of writings and insights into old growth. The funding for Reflections comes from Forest Service research funds and from the Spring Creek Project. The Forest Service's role in supporting this work is part of the agency's responsibility to manage these places dedicated to learning. Emerging themes in the poetry and essays being produced at the HJA include the importance of taking the long view, the critical role of language, especially metaphor, and how the synergy between science and the humanities can help sustain ancient forests. This cross-disciplinary approach can produce a heightened awareness of relationships in the natural world, such as the ecology of fear, and why it matters.[31]

One recent autumn I participated in a program at the HJA. Moore and Goodrich gathered poets, novelists, and essayists from throughout the West to discuss our calling as writers in this time of rapid global change. Over a three-day period, we spent as much time as possible afield, enjoying the luminous late September days. Sitting in a hemlock stand at one of Harmon's log decomposition plots, Swanson taught us how to "read the forest," interpreting stories trees can tell us about earlier events, such as fires, windstorms, and floods. As we shared our writings and impressions about the nature of this forest, Swanson lay down on the moist, mossy ground and gazed up at the canopy far above, his face filled with wonder. Moore sat cross-legged at the base of a hemlock, eyes shut, a blissful expression on her face. Ancient forests have tangible (economic) and intangible (aesthetic) values. That this forest can inspire awe in a scientist who has studied it for decades and can give wings to philosophical ideas about science, the humanities, and relationships in the natural world illustrates its intangible value.

As Goodrich read us Tang Dynasty poems that probed the dark, damp depths of ancient forests, I leaned back against a Douglas-fir and recalled my first trip to the HJA in 2007 as a visiting scholar. By then I was a graduate student in forestry and wildlife with field experience in various other forest types. It was early spring as I walked into an area referred to as Watershed One. Everything was green, even the bark on shrubs, covered with a fine moss. What struck me first about this viscid, verdant place was its multiple spatial scales—from the very large (a 300-foot Douglas-fir) to the minute (droplets of water embedded in the tree's corked, mossed bark). The trees' immensity and my awareness of this forest's age were humbling. Some of the oldest trees in this stand had originated 500 years ago, after a major fire. Sunlight filtered through lacy hemlock boughs, suffusing the understory with a low green light. Other than the song of creek water on stone and the soughing of the wind in the canopy, this place seemed preternaturally silent—I didn't even hear birds or squirrels. I walked on forest trails amid moss-veiled yews, the earth spongy underfoot from millennia of decay. The bones of long-dead trees lay strewn about this landscape, which had been created by the pit-and-mound architecture of decay. Life and death merged seamlessly, with deadfall acting as nurse logs to young saplings, all part of the same continuum.

This place was filled with the contrast of old and new. Artifacts of the state-of-the-art research being conducted here by ecologist Barbara Bond—metal towers, sensors, cables, antennae, and white plastic pipes—rose incongruously from the ground. In one of the most ingenious projects taking place at the HJA, she was documenting how the forest breathes, tracking the miraculous daily movement across this mountainous landscape of what literally amounted to a river of air. Each day the trees breathe, as part of the carbon cycle, pulsing energy through the food web. To learn more about how forests function, Bond followed the nighttime movement of cold air through this landscape with a wireless electronic sensor network designed for this project.

As I walked farther, I wondered how science can do justice to this complex, ancient community. Long-term ecological reflections mean acknowledging that we can't find full answers to ecological questions—including those about trophic cascades—without looking deep and long. LTER sites like the HJA offer places where one can slow down and look at things at the speed of rot. Long-term ecological research entails compiling generations of data, layers of learning, to yield some understanding of the holistic functioning of old-growth forests. Year by year the data accumulate, like leaf litter on the forest floor, each observation contributing to the whole of our knowing, helping us to redefine who we are in relation to the world. Such has been the incredibly fertile research legacy created by Jerry Franklin and carried on by Swanson, Bond, and their many colleagues. Scientists who work in this forest know that there is much more here than we will ever know. Moore, Goodrich, and Swanson are giving them and those involved in the humanities the opportunity to reflect on these matters more deeply.

The Wind River Experimental Forest: Canopy Cascades

My field time as a visiting scholar at the HJA led me to another site to search for trophic cascades in a less accessible part of a forest. Established in 1932, the Wind River Experimental Forest is referred to as the "cradle" of forest research in the Pacific Northwest. This unique facility has a canopy crane, which would enable me to explore food webs in the crowns of old-growth Douglas-firs. It lies in Washington's Cascade Range, in Gifford Pinchot National Forest, eighty miles from the Pacific Ocean. Administered by the Pacific Northwest Research Station in Portland, Oregon, it is one of ten experimental forests within the station's jurisdiction (Oregon, Washington, and Alaska). Additionally, in the early

1900s the Forest Service established research natural areas (RNAs) to represent different types of terrestrial and aquatic ecosystems.[32] The Thornton T. Munger RNA, created in 1926, lies within the Wind River Experimental Forest and contains an old-growth Douglas-fir–western hemlock stand. This cool, moist 500-year-old forest has a mild maritime climate and lies in the transition region between the Coastal Western Hemlock Zone and the Pacific Silver Fir Zone, which occurs above 2,900 feet.[33]

Wind River research concentrated on silviculture through World War II and into the 1970s. In the 1980s the focus shifted to ecosystem studies and relationships of old growth and wildlife habitat. Jerry Franklin, who left the HJA for a sabbatical at Harvard University in the mid-1980s, became the Wind River plant ecologist shortly thereafter. He had spent considerable time in that area as a child growing up in western Washington and thus had a connection to this landscape. In 2007 the National Science Foundation designated the Wind River Experimental Forest one of twenty core sites in the National Ecological Observatory Network (NEON). This program uses cyber-technology, such as environmental sensors, to support interdisciplinary, collaborative research on how environmental changes are affecting forests and biodiversity.

When I arrived at Wind River in late March, several feet of crusty snow covered the ground. It was thirty-five degrees and raining steadily—the sort of rain-on-snow downpour typical at this time of year. I built a fire and made tea in the historic Forest Service cabin where I was housed. As I brought in wood at dusk, I discerned the shadowy forms of elk grazing in an open field (a former tree nursery) across from the cabin and heard a varied thrush's buzz song echo through the forest. Alone in this beautiful, rustic cabin, I spent the evening listening to the drum of rain on the shingle roof and reading about this research facility's more recent history.

In the 1970s HJA botanist William Denison, who insisted that you can't understand forest ecology without looking at what goes on in the treetops, pioneered canopy studies at the HJA. His work inspired vital research by Nalini Nadkarni, Margaret Lowman, and others. In 1980 the Smithsonian Institution installed the first canopy crane in Panama. Prior to that researchers had to use climbing gear (rope) and rock-climbing strategies to access the forest top.

Fourteen years later Jerry Franklin had a crane tower erected at Wind River, and he has been director of this research facility since then, with the assistance of David Shaw during the first decade and Ken Bible more recently. Identical to those used to build skyscrapers, the crane carries an eight-person gondola, has a 279-foot jib, and is as tall as a twenty-five-story building. It offers access to a cylindrical volume of more than 54 million cubic feet, which contains 300 trees. Researchers study such topics as carbon, water, and nutrient cycles; tree physiology and growth; climate variability; the relationship between biodiversity and ecosystem functions; and lichen and fungal ecology.

I rose at dawn the next morning to prepare for my trip into the canopy. The rain had ended overnight, fog blanketed the forest, and the temperature had dropped into the twenties, coating everything with ice. I put on extra wool layers. My companions that day included site director Ken Bible and research scientist Matt Schroeder. They planned to install sensor buttons on trees to gather temperature and humidity data in order to measure habitat variation and its effect on growth and canopy organisms. After donning nylon safety harnesses and hard hats, we stepped into the gondola basket and clipped ourselves to the railing. As the electric-powered crane silently propelled us into the dark, still canopy, we entered another world—one seldom seen at arms' length by humans. More smoothly than any elevator ride I had ever taken, crane operator Mark Creighton expertly lofted us through the lower canopy and into the forest nave and upper canopy. As we rose beyond the cone zone on this quintessential Pacific Northwest spring morning, I could see that the fog was already dissipating. A panoramic view opened of low volcanic mountains densely carpeted with forest, ridgelines fading into shades of distance. The previous night's storm had left everything green and growing, wet on wet, in infinite shades of green. The sun burned through the low clouds and made the icy raindrops that clung to the treetops glitter like crystals.

The crane stopped and left us looking down on the interwoven crowns of ancient trees. My strongest impression was of immense structural and textural diversity. While I had studied canopy structure in silviculture textbooks, which featured tidy diagrams depicting multiple levels of trees, this was the first time I had physically explored this structure. More than simply a collection of treetops, the canopy encompasses all the living (biotic) and nonliving (abiotic) components of trees, such as atmospheric gases and tree litter.[34] Although its medium was air rather than water, the canopy most resembled a coral reef in terms of its structural complexity and the wealth of life it held, including many insects, birds, and lichens. And as I looked I noticed a lacy lichen liberally distributed among the branches in the upper canopy.

The lichen *Lobaria oregana* is composed of lime green, frilled, and veined lettuce-like leaves, which sometimes resemble elephant ears. Primarily growing in ancient forests because of its light and moisture requirements, *Lobaria* accounts for up to half of the foliar biomass in lowland Douglas-firs. It is part of a class of nitrogen-fixing lichens termed *cyanolichens*, which harbor nitrogen-containing bacteria. Rain washes off the bacteria and leaches their nitrogen into the soil, making it available to trees and other plants. *Lobaria* is scarce in young forests, which provide insufficient shade for this genus. Lack of cyanolichens in cutover second-growth forests diminishes the amount of nitrogen cycled through these systems.[35]

Figure 5.1. Old-Growth Douglas-Fir–Hemlock Canopy, Wind River Experimental Forest

The canopy's complex structure creates *microsites*—small, discrete patches of habitat, stratified by elevation and related to the amount of light and moisture that reaches them. Some species prefer certain portions of the canopy. For example, the red tree vole (*Arborimus longicaudus*) uses the lower canopy of the largest trees in old-growth forest. It spends all its life aboveground, where it disperses spores and seeds and provides a valuable food source for predators. Some songbirds, such as the brown creeper (*Certhia americana*), forage selectively. Most of the year this species harvests insects from tree bark in the midcanopy, but in spring it shifts to the high canopy.[36]

Canopies undergo development over time, as is most apparent in old-growth forests, which feature a variety of structures, from the tender green, rapidly growing tops of young hemlocks to the broken, gnarled crowns of ancient Douglas-firs. This parallels the succession pathways discussed earlier in this chapter, caused by competition for light, which drives structural changes in trees, and also by disturbance, which creates openings and enables shade-intolerant species to grow.[37] The canopy before me demonstrated all these processes at once, and it took me the several hours I was up in it to begin to explore its richness. The crane provided the ideal vehicle for this as it glided from tree to tree and level to level.

While Bible continued positioning sensors, I looked for trophic cascades, curious about whether canopies contain some of the food web relationships I had been observing elsewhere. He and Schroeder inspected the tops of hemlocks severely infested with dwarf mistletoe (*Arceuthobium tsugense*), a small, brownish, leafless parasitic plant. When I asked what was causing this outbreak, they said it was due to an exceedingly rare butterfly that lays its eggs only in the tops of mature hemlocks—or, more accurately, due to its lack.

In 1952 Ray Bradbury wrote a short story, "A Sound of Thunder," about how a butterfly had far-reaching effects that rippled through time. In this story a time traveler ventured to the dinosaur era. The scientist who operated the time machine warned him not to disturb anything. As he walked in a Jurassic rain forest the traveler inadvertently killed a butterfly by stepping on it. Upon returning to the present, he found things significantly different—language, political regimes, how the world looked. In science, the notion that a small organism's status and actions can have profound effects came to be known as the *butterfly effect*.[38] As Bible and Schroeder described what amounted to a butterfly effect in this forest, I began to piece together its trophic implications.

The Johnson's hairstreak (*Callophrys johnsoni*), which has been observed in the canopy at Wind River, also commonly known as the mistletoe hairstreak, is a one-and-a-half-inch olive green butterfly marked with brown, orange, or blue spots and a white zigzag on the underside of its wings. Other field marks include double tails and raised dorsal chevrons. According to Robert Michael Pyle, it is the only old-growth obligate butterfly, which means it requires ancient forests to survive. It lays its eggs only on dwarf mistletoe, which the caterpillars eat when they hatch. According to the Xerces Society for Invertebrate Conservation, this butterfly once occurred from southern British Columbia south through eastern and western Washington, Oregon, and western Idaho to central California. It probably existed throughout much of western Washington's old growth prior to 1900. It has severely declined as a result of habitat destruction caused by logging and the spraying of insecticides to kill tussock moths and budworms in conifers. Other than relatively recent sightings in the Wind River Experimental Forest and Olympic National Park, it was last seen in 1969 in King County, Washington. Thus it's likely that when mature hemlocks abounded in the Pacific Northwest, so did the Johnson's hairstreak, keeping the parasitic dwarf mistletoe under control. And as logging depleted these trees, dwarf mistletoe was depleted as well, and this butterfly declined for lack of egg-laying habitat.[39]

The Johnson's hairstreak decline created an ecological chain reaction. Dwarf mistletoe lives in older hemlocks and survives by taking water and nutrients from its host. Too much mistletoe can severely damage a hemlock, impairing the tree's ability to circulate water and nutrients by as much as 50 percent. While most hemlocks survive light infestations, trees with heavier infestations are unable to compete with healthier trees for resources; those whose boles are infested eventually die. A heavy outbreak also greatly reduces a tree's ability to produce cones. Because the Johnson's hairstreak has become so rare, dwarf mistletoe is taking over many of the mature hemlocks in the Pacific Northwest, impairing their health. As these trees continue to decline, they have a lower likelihood of reproducing. Fewer hemlocks mean the decline of other species dependent on their cones for nutrition, such as common crossbills (*Loxia curvirostra*). Existing hemlocks' inability to adequately cycle nutrients reduces resources available for invertebrates and mycorrhizal fungi. And because all things are connected, this negatively affects soil composition, other tree species, understory vegetation, and wildlife.

While the Johnson's hairstreak butterfly can't be considered a keystone species as traditionally defined by Paine (it's not a carnivore), it is a strongly interacting species—one whose presence or absence has tremendous influence. These cascading effects show that food web connections, even involving small organisms such as this brown butterfly, can touch every member of a food web. And while today we know that the "balance" of nature is a myth because nature is filled with uncertainty and unpredictability, removal of a butterfly species can upset or sever evolutionary relationships and cause systems to change across multiple trophic levels.

Riparian Cascades in Temperate Rain Forests

Riparian studies have been part of old-growth forest science since this field's early years. In 1966, inspired by research in other systems, HJA scientist C. David McIntire created an integrated model of the major trophic levels of riparian food webs (e.g., detritivores, carnivorous and herbivorous fish).[40] In chapter 3 we learned about Mary Power's seminal 1980s trophic cascades work in northern California's Eel River. She has also studied how rivers subsidize watersheds in old-growth forests and found a trophic connection between the carbon produced by riparian food webs and insect abundance. In this study, which also took place along the Eel River, carbon indirectly increased numbers of insectivores, including lizards, birds, and bats, a bottom-up effect. In the 1990s HJA hydrologist Steven Wondzell took an even more holistic view, linking rivers and landscapes via nutrient flow in the hyporheic zone and examining the role of disturbance in shaping the resulting energy exchange.[41] Also, since the 1990s hydrologist Julia Jones and her colleagues have been studying differences in water use by hardwoods and conifers and how vegetation age, structure, and species composition affect hydrologic patterns in rivers.[42] In 1997 Robert Naiman, a specialist in Pacific Northwest riparian ecology, added another dimension by suggesting that a combination of herbivory by large mammals and other forms of disturbance, such as floods, shapes riparian areas by modifying vegetation distribution, and that keystone predators may have a role in this. Specifically, he identified the potential indirect trophic effects that wolves and river otters (*Lutra canadensis*) have on plant communities by influencing the abundance, distribution, and browsing choices made by their prey.[43]

Hydrologist Robert Beschta has long been aware of the ecological importance of riparian vegetation, which stabilizes riverbanks and controls the amount of light that reaches streams, thereby regulating water temperature. In 2008 he and William Ripple investigated a potential riparian trophic cascade consisting of wolves, Roosevelt elk, black cottonwoods (*Populus trichocarpa*), and bigleaf maples (*Acer macrophyllum*) on Washington's Olympic Peninsula—a large arm of land west of Seattle, across the Puget Sound. This rain-drenched landscape (about twelve feet of rain falls annually on west-facing valleys, making it the wettest place in the forty-eight coterminous United States) contains the Hoh, Queets, and Quinault rain forests, some of the most productive timberlands in the United States, as well as Olympic National Park. Roads provide access to the park's perimeter but not its primeval interior. The glacier-topped Olympic Mountains rise directly out of the Pacific Ocean and dominate the center of the peninsula, forming a cluster of jagged peaks rimmed by thickly forested foothills and valleys. Out of these mountains flow thirteen salmon-bearing rivers, which descend steeply, incising the timbered foothills as they course swiftly down to fertile valley bottoms, floodplains, and the Pacific Ocean.

As elsewhere, wildlife populations have left their marks on this peninsula's plant communities. Elk abounded here until 1895, when they were overhunted by settlers. In 1905 the federal government implemented a peninsula-wide hunting ban, causing elk numbers to surge, with heavy browsing reported by 1915. By the 1920s humans had extirpated wolves, which exacerbated the irruption, and in 1935 Olaus Murie reported that the excessively high elk population was depleting its food sources, including understory plants.[44] Congress created the park in 1938 to protect old-growth forests and elk habitat; however, without wolves and hunting by humans, intense elk browsing has continued through the present.

Also as elsewhere, wolves have begun returning to Washington, drifting down from southern British Columbia, where they were never fully extirpated, and also dispersing into southeastern Washington from Idaho. In July 2007 a pack denned and produced pups in Washington's Okanagan country in the North Cascades, approximately 100 miles east of the Olympic Peninsula. In 2008 a wolf was spotted in southwestern Washington, on the boundary of the Wind River Experimental Forest. There have been other sightings throughout the Washington Cascades, but as of 2009 no wolves had been reported on the Olympic Peninsula, possibly because of formidable geographic barriers, such as the city of Seattle and Puget Sound.

Beschta and Ripple examined food web effects in Olympic National Park—a protected area—and contrasted that to an area of multiple human land uses outside the park on the Quinault Indian Reservation. The overstory in their study areas consists of red alders, black cottonwoods, bigleaf maples, Sitka spruce (*Picea sitchensis*), western hemlocks, and western red cedars. In response to disturbances such as logging, fire, or flooding, these forests form dense shrub understories. Beschta and Ripple focused on the healthy growth of two common elk foods, black cottonwood and bigleaf maple, along the Hoh and Queets rivers inside the park. For comparison they used a bar on the lower Quinault River, outside the park, that had low elk presence as a result of human activity. At all but the Quinault bar they found a gap in cottonwood and bigleaf maple ages correlated to wolf extirpation, consistent with a trophic cascades hypothesis in which the abundance of a keystone predator (wolf) influences the abundance of its primary prey (elk), which influences plant communities (cottonwoods and maples). They attributed cottonwood and bigleaf maple recruitment on the lower Quinault to human activity and land uses such as hunting.

Franklin and other scientists agree that elk are a major ecological force shaping vegetation patterns in valley bottom old-growth forests in Olympic National Park. This has been documented in studies by ecologists and is very obvi-

Figure 5.2. Ungulate Exclosure, Olympic National Park

ous in two exclosures Franklin was instrumental in establishing in the South Fork of the Hoh River in 1980. In the mid-1990s Andrea Woodward and Ed Schreiner found that in these exclosures, and in others outside the park, elk influence shrub and tree recruitment by feeding preferentially on western hemlocks and western red cedars and by avoiding Sitka spruce, an unpalatable species. Exclusion of elk decreased the amount of grass and increased tree density. Shrubs inside exclosures had higher diversity and included palatable browse species such as salmonberry (*Rubus spectabilis*) and huckleberry (*Vaccinium* spp.), which had disappeared elsewhere as a result of elk preferential browsing.[45]

While as of 2009 wolves hadn't reached the Olympic Peninsula, Franklin predicts that if they return to this ecosystem, wolves will reduce elk density and impacts on understory communities. However, Franklin and Olympic National Park ecologists have an alternative perspective on the value of using cottonwoods as an indicator of these dynamics. Although the cottonwood represents an ecologically important component of riparian plant communities in general, in Olympic National Park it makes up an insignificant proportion of the hardwood community (less than 1 percent). Franklin suggests that studies based on its recruitment may be biased by failing to focus on species with greater importance in this system, such as red alder. But elk do not seem to feed much, if at all, on alder because it produces compounds that render it fairly unpalatable. This illustrates that relationships between elk and various browse species create complex community dynamics that bear further study.[46]

According to Beschta and Ripple, trophic cascades have influenced hydrology inside and outside the park. They found river channels inside the park more braided (37 percent) than outside the park (3 percent), with significantly more bare soil. They noted that these reaches historically may have been single channels, and they linked the shift to a braided condition to intense elk herbivory in riparian areas. They interpreted braiding as evidence of a degraded stream, caused by elk removing stream bank vegetation and thereby making the earth less stable alongside a river. Meanwhile, they attributed the lesser extent of braiding outside the park to human activity, such as hunting, that reduced elk herbivory on cottonwoods and bigleaf maples.[47]

Like everything else in nature, riparian ecology is complex and multicausal. Naiman suggests that while herbivory has a strong effect on stream banks by influencing plant growth and distribution, other forms of disturbance may have an equal role in shaping some of these dynamics. The biophysical processes at work in Pacific Northwest floodplain systems fundamentally shape ecological community dynamics. Debris flows, characteristic of healthy riparian systems, scour and widen streams and send big logs hurtling downriver. These events undercut stream banks, remove vegetation, and create changes in plant communities, which include succession and production of trees. Naiman proposes that the missing cottonwood age classes Beschta and Ripple observed could also have been influenced by debris flows. Additionally, Naiman and Swanson assert that braided streams are not necessarily evidence of riparian degradation. A braided river contains a network of narrow channels separated by small, temporary islands or gravel bars. Braiding occurs naturally in rivers that have a high slope and carry a large sediment load and also where rivers dramatically decrease in width as they reach floodplains, as is the case in Olympic National Park. In this park thick alder stands grow along many stream banks, helping to hold the soil in place. Naiman and his colleagues suggest that reducing large predators, especially wolves, may have resulted in significantly decreased recruitment of browsable plants that can stabilize stream banks in the Pacific Northwest; however, a braided river may not be primarily the result of this. Although it's very likely that wolves could have a strong top-down effect on this ecosystem, as elsewhere, the complexity of these interactions invites deeper inquiry.[48] Given Washington's recolonizing wolf population, we may someday have the opportunity to test these dynamics more closely.

Wind River Wolf Cascades?

At the Wind River Experimental Forest, I went for a walk in the Thornton T. Munger Research Natural Area old-growth stand with Ken Bible to examine elk herbivory. Franklin had suggested that elk, which have a high population here, have been attracted by human land use practices that create forest openings, such as timber harvesting and the abandoned tree nursery across from the cabin where I stayed. Additionally, an elk reintroduction conducted locally in the previous century, when elk numbers had reached a low point, created an unnatural situation. Nevertheless, ungulate impacts have been similar to those observed in other wolfless places, such as the upper Midwest, where intense herbivory eliminated *Trillium* from the understory.[49]

As we walked, Bible showed me abundant elk beds, tracks, and droppings that went deep into the forest, several hundred yards in from the edge. He pointed out salal (*Gaultheria shallon*), Oregon grape (*Mahonia repens*), and huckleberry pruned low to the ground and stripped of their leaves by elk. He and Franklin have installed an exclosure to begin to track the effects of herbivory here, such as changes in plant species composition. Western red cedar may provide an example of these effects. Bible explained how virtually none grew in the understory, even though this species historically had been present. Given that it is a highly palatable elk food, its absence may be related to intense herbivory. However, this pattern may soon change.[50] If the recolonizing wolf population develops at a rate similar to that in other places, such as Oregon and Montana, an ecologically effective population may become established at Wind River before long. When it does, researchers will document any changes in the forest trees and shrubs in response to a release from herbivory and other changes in this food web's dynamics.[51]

The ecological effects of wolves in Pacific Northwest rain forests may be markedly different from those in simpler systems, such as Yellowstone National Park. According to Franklin, places where concentrations of ungulates exist in this region are quite limited in spatial extent. He has found that ungulate distribution tends to reflect human settlement and activities (e.g., logging) or large and persistent natural disturbances (e.g., Mount St. Helens), which create openings in old growth that increase sprouting of tender shoots of shrubs and trees. Hence, any trophic cascades study involving wolves, elk, and plant communities should ideally consider the dominant forest landscape condition (prelogging, pre–volcanic eruption), along with the areas where we expect to see some effects of the wolf's return.

TROPHIC CASCADES in highly speciose systems are not as easy to quantify as those in simpler systems. In old-growth rain forests, as elsewhere, top-down energy combines with bottom-up effects, which include nutrient flow and disturbance. Nevertheless, trophic cascades ideas about the value of keystone predation are meaningful here. While ancient forests harbor a wealth of life, they may lack some of the highly interacting species once present, such as gomphotheres, jaguars, or wolves. We have seen how in the tropics some can be considered empty forests, with significantly altered trophic interactions. In temperate old growth, recolonizing wolves will indirectly change species assemblages as populations of trees respond to changes in herbivory, but bottom-up factors, such as disturbances (fire, floods), may interact with these effects, dampening them. Regardless of the strength or scope of keystone effects in old-growth rain forests, becoming aware of missing trophic relationships is an essential first step toward restoring ecosystems. In the next chapter we will learn more about how these relationships influence biodiversity.

PART TWO

Mending the Web

CHAPTER SIX

All Our Relations: Trophic Cascades and the Diversity of Life

At every turn in my research path the wolves I study have taught me lessons about biodiversity, many resulting from my direct experiences with the powerful effects they wield in communities. At dawn one spring I sat cross-legged on the ground with my field crew under some scorched aspens, remnants of a 1988 wildfire, on the edge of a large wet meadow in Glacier National Park. This meadow was located in the North Fork, which had a very high density of wolves, one some biologists believed to be at carrying capacity. Mist fingered the meadow, lending an ethereal quality to the landscape. I was doing a point count, which involved recording all the birds I detected visually or by song during a ten-minute observation period. I was measuring songbird diversity in areas of high, medium, and low wolf density. We sat in silence, almost perfectly still, to avoid disturbing the birds. This peaceful time in the rosy dawn light afforded me the opportunity to witness wildlife interactions seldom observed by humans.

The birds had been active that morning, some of them species of special concern: olive-sided flycatcher, black-backed woodpecker, American redstart. A neon flock of mountain bluebirds flicked through the meadow, gleaning insects from the air. From the willows came the lilting melodies of six warbler species. Beneath their songs I detected four kinds of flycatcher, and three woodpecker species providing a rhythm track with their staccato drums and high-pitched jungle bird cries. Some of these species were declining throughout their range because of human modification of ecosystems. Here the web of life thrummed like a symphony, with all the players present.

Sudden movement erupted from the western end of the meadow. In the luminous dawn light I made out a female moose, running the way animals do when fleeing for their lives, legs fully extended, like those of a racehorse. Across the meadow from us my daughter Bianca, who was part of my field crew that morning, pointed out two pale gray shapes highlighted by the rising sun. Wolves chasing the moose in calving season. The moose cow neared, becoming very agitated when she spotted us. She began to spin, buck, and snort, close enough that we could see the whites of her eyes, flecks of foam on her muzzle, and her fear-erected nape hairs. The wolves had probably taken her calf and were now after her. To make matters worse, she had been planning to escape by running to the edge of the meadow, only to find us in her path, a researcher with an aluminum clipboard and three field technicians, all sitting there stunned.

She began to swing her legs. I recalled accounts of moose killing humans with their hooves. She could be on top of us in seconds, and there was nothing we could do. The spindly aspens provided no cover against an enraged mother who stood six feet tall at the shoulder. As I considered our predicament, she abruptly bolted, vanishing into the aspens at the meadow's eastern end, the wolves closing in. The birds resumed their songs. We sat there, hearts beating hard, understanding clearly that there were primal forces beyond our human ability to control, and survival and predation were everything.

Months later, my data for this amazing location in the North Fork revealed many more songbird species than had been present at my research site on the eastern side of Glacier National Park, which has a very low wolf population. Even after accounting for all possible influences, wolves appeared to be driving this effect by indirectly creating bird habitat. And the story of that calf's death and the moose mother's grief dramatized my results; keystone predation was shaping so much of the pattern at this site, even the rich layers of birdsong.

What Is Biodiversity?

Since the mid-1990s conservation biologists have acknowledged that one of the leading effects of trophic cascades is an increase in biodiversity, which Aldo Leopold observed in the 1930s and 1940s. Thus it follows that restoring keystone predators may help conserve biological diversity in many systems—from the deep oceans to the Rocky Mountains. Any discussion about the mechanics of how keystone species affect biodiversity benefits from defining some of the science that supports this idea.

Conservation biology is the science of analyzing and protecting the earth's biological diversity. It addresses human effects on the environment, with the objective of sustaining the web of life.[1] A relatively new science, it was founded in the mid-1980s by Michael Soulé, Bruce Wilcox, David Ehrenfeld, Peter Brussard, and other scientists concerned about widespread habitat degradation and species extinctions. Government agencies were doing little to stop what these scientists perceived as an ecological crisis in the making. Soulé and his colleagues hoped to shed light on key ecological interactions and create conservation tools that could be applied by agencies and landowners to restore and protect ecosystems. This new science was controversial at first but rapidly gained acceptance, perhaps because of the exigencies it addressed.

Ecologists Walter Rosen and Elliott Norse created the term *biodiversity* in 1985 from the phrase *biological diversity*, which refers to number of species.[2] It is a rich term, rife with meaning. Conservation biologists Reed Noss and Allen Cooperrider define it as "the variety of life and its processes. It includes the variety of living organisms, the genetic differences among them, the communities and ecosystems in which they occur, and the ecological evolutionary processes that keep them functioning, yet ever changing and adapting."[3]

The *species* is the fundamental unit of biodiversity, defined by Edward O. Wilson as "a population whose members are able to interbreed freely under natural conditions." Species matter for evolutionary reasons, to ensure that life in all its forms continues on this planet, but also for all the known and unknown ecosystem benefits we can derive from them. However, Robert Paine notes that not all species are equal, referring to Charles Elton's food pyramid, in which top predators, much lower in abundance than their prey, are therefore far more susceptible to extinction. They suffer first when an ecosystem starts to erode from the bottom up because of climate variability or human-caused changes, including addition of pollutants and toxins such as lead, mercury, and DDT.[4] They also exert the strongest influence on a food web, from the top down.

The role top predators play in a community illustrates the complexity of biodiversity, which is composed of hierarchical levels: genetic, species-population, and community-ecosystem. Using the sea otter as an example, this species' unique set of genes represents the *genetic* level of biodiversity. Genetic diversity enables species to adapt to disturbance and change. For instance, if a lethal disease breaks out in a sea otter community, diverse genetic makeup may allow some individuals to survive, thus enabling the persistence of the community. The *species-population* level would include sea otters living on one of the Aleutian Islands. Also called a *deme*, this level of diversity helps a population adapt to change. Finally, the *community-ecosystem* level of biodiversity would encompass all sea otters living on all of the Aleutian Islands and all other organisms living on or near the islands that make up this environment. This includes *structure* within ecosystems—for example, where sea otters fit into a food web, their role within this framework, and how they interact with other organisms, such as their sea urchin prey. It also includes structure outside of ecosystems, such as how ecosystems link to one another in a geographic region. Thus, in the otter example, ecosystem biodiversity results in many different types of marine ecosystems, including coastal and intertidal. And it creates resilience in dealing with harmful disturbances such as oil spills.[5] All of this suggests that to fully understand and conserve biodiversity, it's necessary to consider whole ecosystems, not just species.

Every ecosystem is biologically diverse, some more so than others. Even a highly modified ecosystem, such as a clear-cut forest, can contain much diversity, some of it hidden (e.g., organisms that live in the soil). Ecosystems such as

Mount St. Helens in the years after its 1980 eruption have shown very high biodiversity over time as they go through the early seral stages of recolonization of organisms and succession. Indeed, a disturbed community's biodiversity sometimes can surpass that of some systems in the later stages of development, such as a mature forest. But an old-growth forest (one 200 years old or older) is likely to contain biodiversity elements and processes absent from a younger forest, as we have seen.

If biodiversity can be promoted by a variety of means, such as disturbance, why are keystone predators considered so important? Most conservation biologists agree that this is because keystone presence supports aspects of diversity otherwise not present or fully functional in ecosystems, by indirectly creating habitat for community components such as birds and fish. In this chapter I discuss structural biodiversity patterns created by keystone predators across a variety of ecosystem types.

The Naturalist

Harvard University evolutionary ecologist Edward O. Wilson, arguably one of the great scientific thinkers of the twentieth century, forged early theoretical concepts about how biodiversity works. He was among the first to note the relevance of trophic cascades in maintaining biological diversity in ecosystems and the first to tie food web effects that help sustain diversity to benefits humans might derive from ecosystems. In 1999, in his book *The Diversity of Life*, he commented, "The loss of a keystone species is like a drill accidentally striking a powerline. It causes lights to go out all over."[6] Any narrative about biodiversity should include the story of his work.

Had you visited Paradise Beach, Florida, in 1936, you might have seen a seven-year-old boy sitting alone on a pier, gazing intently for hours at the animal life teeming beneath the water's surface: jellyfish, needlefish, blue crabs, and stingrays. He was blinded in one eye that year, during an unfortunate fishing accident that occurred when he yanked too hard on his line and a fish flew out of the water and struck him in the face. He lost stereoscopic vision but was left with exceptionally sharp vision in his good eye—so sharp that he could make out the fine hairs on an insect's body. From his start in life as a deeply curious and precocious only child who enjoyed grubbing in dirt and exploring swamps, Wilson proceeded to give full rein to his all-consuming fascination with the natural world. He dreamed of becoming a professional field biologist. His poor vision made him a mediocre birder but opened the insect world to him. By his midtwenties his dream had become a reality. Through hard work and multiple fellowships he had obtained a master's degree at the University of Tennessee and a Harvard PhD, working in the field of myrmecology (ants), which had by then become his primary passion.[7]

This passion took Wilson to remote tropical islands in the South Pacific, where he combed rain forests and mountaintops for rare and new ant species. A patient, sedulous field biologist, he worked in Melanesia, south of the equator, exploring islands that were in effect self-contained worlds. In the process he began to develop daring ideas about species diversity and habitat. By the early 1960s he had become a tenured professor at Harvard and had partnered with Robert MacArthur to study the relationship between the size of islands and the number of species they could sustain.

Like Elton and Leopold, the team of MacArthur and Wilson was a natural partnering of two great intellects with compatible interests and temperaments. A consummate field ecologist and synthesist who had big ideas, Wilson perfectly complemented MacArthur's mathematical brilliance, attention to detail, and theoretical genius. In the early 1960s they would meet at MacArthur's home in Marlboro, Vermont, where they had intense conversations about ecology, biology, and *biogeography*—the geographic distribution of plants and animals. Wilson was deeply interested in the patterns of species diversity he'd found on islands and saw biogeography as an important vanguard. Years before, when graphing his Melanesian ant data, he had created something called a *species-area curve.* The line on the graph he produced depicted the correlation among a group of different-sized islands between area and number of resident species. Wilson's graph suggested that the larger the island, the more species it could sustain. Smaller islands contained fewer species. In practice Wilson found that a 50 percent decrease in island size led to a tenfold decrease in species. Thus an island that measured 50 acres would have ten times fewer species than an island that measured 100 acres, and ten islands that measured 10 acres would not have among them nearly the number of species that a 100-acre island would.

MacArthur and Wilson sought to refine this relationship, to find patterns in what Charles Darwin saw as the tangled bank of the natural world that could be quantified in an equilibrium model for islands. They were young, audacious, and full of exciting ideas. What became known as their theory of island biogeography, essentially a formula for species loss and gain, created a furor in the scientific world. Its beauty lies in how it also applies to terrestrial systems isolated by natural or human elements. Contemporary examples include the sky islands in the southwestern United States, or habitat patches in Southern California surrounded by a "sea" of desert or urban area. MacArthur and Wilson's idea debuted as a hypothesis in a journal article, which they quickly fleshed out into a full-blown theory, published in book form in 1967.[8]

The theory goes like this: Let's say you have a brand-new island that has no species. Gradually it becomes colonized by organisms that come in from the ocean and on the wind from other landforms. The probability that a given species will become extinct increases as more species arrive. As the island fills up, the rate of extinction rises and the rate of immigration drops. These processes can be depicted as two separate lines on one graph. As one goes up the other goes down. Where the extinction and colonization lines cross on the graph represents the point of dynamic equilibrium in species diversity. It sounds simple, but the mathematics behind this theory involve virtuoso riffs of differential calculus. To test their hypothesis, MacArthur and Wilson used data from Krakatau, a small island located between Sumatra and Java where all life had been wiped out by a series of massive volcanic eruptions in 1883. Within a year of this event scientists visited the blast zone to document life's return to this island, and over several decades they created a record of species recolonization. MacArthur and Wilson focused on the bird data, which showed that by 1927 the island held twenty-seven bird species and that by 1934 this number had not increased, suggesting the bird population had reached a dynamic equilibrium of species.[9]

Not everyone accepted MacArthur and Wilson's theory. For one thing, the Krakatau data were spotty and of questionable quality and didn't quite fit their formula. Undaunted, Wilson proceeded to test the theory experimentally with his student Daniel Simberloff. Finding an island totally devoid of life was not easy. Wilson and Simberloff ended up creating their own such ecosystems in the Florida Keys. They fumigated certain islands, to wipe them clean of animal life, and then documented species colonization and subsequent extinction rates. In 1969 Simberloff published their results, which met the predictions of the theory of island biogeography.[10]

More than forty years since its publication, *The Theory of Island Biogeography* is considered by many one of the pillars of ecology. MacArthur tragically died in 1972 of renal cancer, leaving Wilson to continue on his own. He went on to pursue a broad range of work, in the process winning two Pulitzer Prizes and a host of other honors. Today he is regarded as a doyen of evolutionary biology and biodiversity science.

In the 1970s Simberloff and other scientists challenged the theory of island biogeography's applicability, finding cases where the species-area relationship didn't hold. For example, they identified a relationship between habitat variety (called *heterogeneity*) and number of species. The more varied the habitat, the more diverse the species an island could hold. Thus small islands that had several different types of habitat could sustain a greater number of species than larger islands that had only one type of habitat. Simberloff tested this experimentally on the very islands where he and Wilson had done their research in the late 1960s and found a correlation between habitat heterogeneity and species diversity. In practice today conservation biologists use both principles (island biogeography and habitat heterogeneity) when designing reserves to support biodiversity.[11]

Endangered Ecosystems

Predation by large carnivores through cascading effects across progressively lower trophic levels (e.g., prey to plant communities to other faunal species living in the plant communities) may be critical to maintain biodiversity. Even as we are striving to expand our knowledge of species interactions, biodiversity is decreasing at an alarming rate. In 1993 Wilson estimated that we are losing as many as 30,000 species per year—a sobering three species per hour. Extinction is forever, but it is not new to this planet. Five known major waves of extinction have passed over the earth in the past 440 million years. We are currently in the sixth extinction wave, this one caused by human impacts on species and ecosystems.[12] Some scientists think this last wave began with the advent of agriculture, which allowed more people to occupy more areas.

Additionally, three types of extinction occur: (1) global extinction, the disappearance of a species from its entire range; (2) local extinction, the disappearance of a species from a portion of its range; and (3) ecological extinction, the reduction of a species in number to a level below which it is no longer ecologically effective in terms of its interactions with other species.[13] Again using as an example the sea otter, which once ranged across the Pacific Rim from northern Japan to central Baja California, global extinction would involve disappearance of this species from its entire range. Sea otters were hunted intensively for their pelts during the Pacific fur trade in the mid-1880s. By the early twentieth century this species had been eliminated from most of its range, thus experiencing many local extinctions. Some of the colonies that remained may have contained enough sea otters to consume sufficient numbers of sea urchins to drive a trophic cascade. On the other hand, on some rocky coves where only a handful of otters remained, they may not have been sufficient in number to have much impact on the food web and therefore could be considered ecologically extinct.

The United States hosts 100,000 known native terrestrial and freshwater species of vertebrate and invertebrate animals and plants. While conservation biologists don't have much data for many of these species, what we do know suggests that of them 1 percent have become extinct in the past two or three centuries, 16 percent are in danger of extinction, and 15 percent are considered vulnerable. This means that one-third of the species in the United States are of conservation concern, mostly as a result of human modification of ecosystems. It's time to take a hard look at the kind of world we are creating.[14]

The game Jenga involves removing wooden blocks one at a time from a tower without causing it to collapse. As you remove them the tower starts to teeter, loosening some of the blocks and making them easier to remove. Eventually, of course, the whole system collapses. Is extinction like a game of Jenga, where you remove more and more species but the ecosystem still stands until you remove a fateful species and the whole thing comes crashing down? As Wilson put it, if enough species are extinguished, will the ecosystem collapse, and will extinction of most species soon follow?

Let's play an ecological game of Jenga to see how this might work. Let's say an aspen community contains fifty species of songbirds, twelve species of woody shrubs, one species of mesopredator (coyotes, in this system), beavers, three species of weasels, and four species of rodents. The birds include three species of chickadees, six species of warblers, four species of flycatchers, and three kinds of raptors. Large predators include wolves, grizzly bears, cougars, and black bears. Ungulates include moose, elk, and white-tailed deer. All have reached dynamic equilibrium in their populations, which means their numbers fluctuate within a historical range of variability, not going very low or very high. This particular aspen stand is large—covering ten square miles, what is referred to as an aspen parkland. And let's say that, just as in a game of Jenga, you start removing species.

You remove the black-capped chickadees first and not much happens; the system continues functioning more or less as usual. The number of chestnut-backed chickadees increases, to fill the gap left by the black-caps. Next you remove short-tailed weasels. The mouse population initially goes up, but then the coyotes start eating the surplus mice, as do the northern harriers, and mouse numbers go back down to the level where they were before you removed the weasels. When you remove willows, things start to get dicey. Willow flycatchers leave the system, as do American redstart warblers, species dependent on willows for nesting habitat. But the other flycatcher species surge in number because they have more insects to eat, and the system continues to function more or less as usual. A casual observer wouldn't notice too much missing. You continue removing species, and each time you do you tug on the other species because they are all connected, maybe not directly but connected just the same, indirectly, via food web pathways. You remove the white-tailed deer and the ungulates carry on, business as usual, filling in the gap by producing more elk and moose. You get daring and remove more shrubs. More songbird species leave. They don't come back.

Next you turn to the predators. Deciding to go for broke, you remove wolves. At first nothing happens. Cougars and bears have more food. The ungulates seem content and far more relaxed. The remaining birds are fine with all this. In fact, initially the systems feels a lot more peaceful. But then, gradually at first, the whole system starts to transform into a recognizably different kind of place. Ungulate numbers go way up because cougars and bears can't put much of a dent in them. The bears sleep half the year, and when they are awake they're just as apt to eat berries as elk, so they're not much help. And cougars, because of their solitary hunting habits and low population densities, have a much lower impact on ungulates in this system than wolves did. The ungulates start running out of food. New trees and shrubs keep sprouting, but ungulates keep them hedged down to a foot or so. Birds that nest and feed in the aspen canopy's midstory leave. With sustained intense herbivory, shrubs and wildflowers disappear. Most butterflies leave too, their host plants eliminated by herbivory. And in less than a decade you are left with an impoverished aspen stand that consists of maybe a third of the species it once held, with very few songbirds or butterflies, no new trees growing into the forest canopy, and a barren understory. Some would call this process ecosystem collapse. Others would say that's too strong a term because the ecosystem is still functioning, albeit limpingly. Others would call it a reverse trophic cascade, ecosystem decay, or ecosystem simplification. Call it what you will, you end up with a degraded ecosystem that's a specter of its former self, in which its fundamental function—energy cycling—may be profoundly diminished.

Biodiversity loss has become a crucial issue in the past two decades as human-caused ecosystem modifications have continued to precipitate a hemorrhage of extinction. Human actions that create islands of habitat in a sea of development put island biogeographic theory to work on the mainland. Secondary extinctions, caused by habitat isolation, are perhaps the most insidious because these effects are not obvious at first and operate on time lags. This means that often by the time we become aware of a problem, it is too late. In connected habitats, trophic cascades increase plant growth; the resulting energy surge drives nutrients across ecosystems, with significant effects that include an increase in biodiversity. Reed Noss and colleagues identified ecosystems in the United States threatened by species loss, mostly because of degradation of habitat by human land use. Many of these systems are ones in which trophic cascades have the potential to increase species richness. These threatened landscapes include aspen parkland, which has experienced localized declines ranging from 22 to 78 percent throughout the United States, and salt marshes, which have declined by 68 percent. While some of these declines can be attributed to bottom-up effects such as climate variability, they are also driven by an interaction with top-down factors, which involves the truncated food chain described in our ecological game of Jenga. Hotspots of biodiversity loss worldwide include native grasslands, coral reefs, tropical rain forests, riparian systems, and coastal forests.[15]

Protected lands, such as national parks, provide examples of how we can conserve endangered species and endangered landscapes. Yet these areas constitute only 3 percent of the world's surface, and until recently few were established with biodiversity conservation as their guiding purpose. Increasing the number of nature reserves will help, but that alone will not do much to protect biodiversity, which includes species within reserves that leave for other habitats for portions of the year, such as neotropical migratory birds and migratory ungulates and their predators.[16] An ecosystem-scale approach that incorporates public and private lands, protected lands, and areas of multiple human land uses, with a keystone predator as the focal species, provides one of the soundest ways to restore and conserve biodiversity.

Why Does Biodiversity Matter?

One could argue that in a system with multiple species of garter snakes, if one became extinct this would have no appreciable effect on the environment. So why does biodiversity matter? Why do so many seemingly redundant species exist? The Jenga example begins to arrive at the answer: loss of species eventually creates loss of ecosystem structure. But there is more to it than that. In his classic paper "Homage to Santa Rosalia," evolutionary ecologist G. Evelyn Hutchinson explored this question. Inspired by Elton's studies on food webs, he concluded that so many different species exist at least partly because trophic complexity makes a community more stable.[17]

Wilson pointed out that we are tied in myriad ways to the millions of the earth's species. If we sever enough of these connections, then *we* can become threatened as a species. Humanity's well-being profoundly depends on biodiversity. No species is truly redundant because all contribute to an ecosystem's ability to function efficiently. In particular, truncated trophic cascades, via loss of keystone species, can result in significant reductions in species diversity. The resulting change in food web structure can seriously affect the ability of ecosystems to provide the ecosystem services on which all life depends.[18]

Ecosystem services related to biodiversity include clean air and water, soil formation, filtration of pollution, nutrient cycling, climate regulation, and recovery from disturbances. Generally, the more diverse the ecosystem, the better it can absorb the carbon dioxide and nitrogen produced by human activity and urbanization, and the more resilient it becomes to perturbations.[19] A system kept productive by multiple plant species has less risk of failing because most communities have apparent redundancies built into them. This means that in the event of a disturbance that eliminates one of these species, another can fill its niche. As species richness drops, ecosystem resilience declines, as does food web efficiency. Additionally, specific ecosystem services may be tied to a particular species found in a particular location, thereby making these losses more significant in terms of ecosystem function. Although we don't know why we need each species of chickadee or spider, we are likely to find out later, when it's too late, why it's wise to save all the pieces, including those.[20]

Other services derived from biodiversity include wood and other products, food, and pharmaceuticals. According to Wilson, 30,000 species of edible plants exist, with 7,000 of them cultivated for food throughout history. Currently 41 percent of prescription drugs used in the United States come from living organisms, and more than 70 percent of anticancer drugs come from rain forest plants. The term *bioprospecting* refers to the exploration of biodiversity to find new medicines. We have barely begun to tap the rich store of possible pharmaceuticals nature can provide. Many of these unmined riches, particularly those endemic to tropical rain forests, are in danger of becoming extinct before we are aware that they exist.[21] Additionally, humans receive many recreational, psychological, and spiritual services from nature.

Keystone Species and Biodiversity

Biodiversity can be increased by disturbance such as fire or flooding, and grazing by large herbivores, such as bison. It is also driven by relationships. These include *symbiosis*, broadly defined as a close association of two or more species. Three types of symbiosis increase biodiversity: *Parasitism* consists of predation in which prey are consumed in small units. Harm is done, but it doesn't result in death. *Commensalism* refers to an interaction between two species in which one experiences a positive effect and the other experiences no effect. An example would be the mites that live on lichens in a rain forest. *Mutualism* involves a relationship in which both organisms benefit, frequently one of complete dependence. Wilson referred to this as an "intimate coexistence" of two species. Examples include flower pollination by a unique species of butterfly, or mycorrhizal fungi living among the roots of a Douglas-fir, helping it absorb nutrients from the soil.[22]

John Terborgh believes that to perpetuate biodiversity over the long run, we must do more besides set aside parks and reserves. This requires preserving the interactions among organisms that have maintained biodiversity for millennia—things such as predation, pollination, symbiosis in all its forms, seed dispersal, and herbivory. And among the most critical of these interactions are those at the top of the food chain.[23] According to Michael Soulé, "in the long run, without restoration of top predators, we'll never be able to protect most biodiversity."[24] Wildlife ecologist and policy expert Hal Salwasser agrees and suggests an even broader approach:

> We must also provide for species to have the ability to move in response to environmental change and to reform into novel communities and ecosystems as occurred repeatedly during the glacial cycles of the past 2.75 million years. And, in the longest run, we need to acknowledge that mega disturbances beyond human control will

reset the whole biosphere and initiate new pulses of extinction and evolution as has happened at least five times since life began on Earth.[25]

One of the most important things related to an increase in biodiversity is predation by keystone species. Robert Paine created the keystone concept to refer to a dominant predator that consumes and controls the abundance of a particular prey species, and a prey species that competes with and excludes other species in its trophic class from the community. The idea is that when the keystone is removed, ecosystem structure and species composition become simplified because of the effects of prey on other species. The keystone term was intended to convey a sense of nature's dynamic fragility *and* tenacity and the unexpected consequences of removing (or adding) species. The keystone concept's attractiveness lies in its focus on the entire ecological assemblage and the recognition that one species can have a disproportionate effect. As Wilson explained, "in communities there are little players and big players, and the biggest players are the keystone species." Thus the keystone concept describes a species whose removal causes substantial changes in a community.[26]

We have become well acquainted with the profound effects that keystone species, such as sea otters (*Enhydra lutris*), sharks (*Carcharhinus* sp.), and wolves, have on ecosystems. In all cases research has shown that removing them has caused significant reduction of biodiversity across multiple food web levels. However, their removal also makes ecosystems vulnerable to invasion by nonnative species, which compete with native species for resources. Invaders thrive because their new environment lacks ecological controls (e.g., predators, competitors) that in their native environments normally would keep them in check.

We saw in chapter 3 the results of keystone species removal in Maryland's Chesapeake Bay, where in the past 200 years humans overfished sharks to the point of extirpation. But there is far more to this story. After the sharks disappeared, we also overfished large predatory fishes such as the Atlantic cod (*Gadus morhua*). Blue crabs (*Callinectes sapidus*), once preyed upon by the now extirpated guild of higher-order predators, irrupted and then were overharvested by humans. The resulting simplified trophic system has become far less resilient and more susceptible to species invasions and bottom-up forces. This is compounded by the fact that Chesapeake Bay is an *estuary*, which receives salt water from the Atlantic Ocean and freshwater from more than fifty rivers. Estuaries are essentially large mixing areas where nutrients vital to plant and animal life drain from the land, uniting the chemical, physical, and biological elements of river and sea to create rich but delicate ecosystems.

Waste produced by urbanization and industry around Chesapeake Bay has drained into this estuary, increasing sedimentation and pollution and introducing excessive nutrients, such as phosphorus (from fertilizer use), leading to eutrophication. Additionally, zebra mussels (*Dreissena polymorpha*) pose a serious threat. Native to the Caspian and Black seas in eastern Europe, this species inadvertently entered the United States in 1988 via ballast water from a European ship. By 1989 the mussels had spread to most of the major rivers and lakes in eastern North America. In Chesapeake Bay, zebra mussels reduce the nutrients available to native species. Lack of a full predator guild impairs this system's ability to fend off these invaders. While restoring keystone species (e.g., sharks) would begin to repair some of the damage, this is no longer a simple matter because of all the other trophic levels (cod, crabs, kelp) that have been impacted. Chesapeake Bay provides a compelling example of the ecosystem vulnerability that can result from systematic removal of predators in a multipredator system.[27]

Habitat Enrichment, Diversity Gradients, and Tropical Trophics

If you start out at either of the earth's poles, diversity increases as you move toward the equator, with much of it sequestered in tropical rain forests and coral reefs.[28] Scientists have tested various hypotheses to explain why these areas are so richly endowed with species. Today most agree it has to do with solar energy and climate stability. All other things being equal, the more sunlight, the greater the diversity (and biomass). The tropics' stable climate and rich resources enable plants to grow more and also supports more animal species. Cold winters beyond the tropics cause photosynthesis to slow, and eventually come to a halt, as one ventures north. Thus in tropical systems bottom-up effects, such as added energy due to the longer growing season, often can prevail. We have seen how in systems with many species, such as coral reefs, trophic cascades may become trophic trickles, petering out as top-down forces become weakened between levels. This occurs because the tropics' high energy availability, in the form of plant materials, enables multiple species to exist in each class (e.g., medium-sized herbivores such as monkeys) and also creates more omnivory than in other systems with fewer resources. Thus animals that eat both meat and plants can cause trophic cascades to leave less distinct signatures in this system.[29]

As one approaches the tropics, ecosystems are characterized by gradients not only in diversity but also in plant defenses, herbivory, and predator activity, which are interconnected forces. This may result in a stronger top-down and bottom-up feedback loop, which can work like this: In the tropics, plants respond to a higher diversity of herbivores by developing stronger defenses in the form of noxious compounds. This creates a bottom-up effect. In the predator guild, ants provide an example of how predator diversity is heightened in the tropics. Wilson found 275 species of ants in a twenty-acre tropical plot, compared with 40 species in a similar-sized plot at temperate latitudes. Furthermore, ants are categorized as predators because they eat other animals, but they also eat vegetation.[30] Their omnivory can weaken top-down trophic cascades.

These examples show that tropical trophic cascades are more complex, characterized by greater herbivory and intraguild predation (ant species preying on other ant species), longer trophic chains (five-level trophic chains, versus the simpler three-level trophic chains found in places such as Yellowstone National Park), and less predictable indirect effects of predation. However, scientists have found many tropical food webs that show evidence of strong top-down effects, such as in Terborgh's studies in Panama and Venezuela.

Mitakuye Oyasin

The Lakota Indians have a powerful prayer, "Mitakuye Oyasin," which means "All My Relations" or "We Are All Related." It is used to invoke blessings on all the earth's living beings and to honor all of life. It also fosters awareness of how all life is connected—as part of a vast web. Keystone predators are not a magical solution that can instantly repair damaged landscapes, owing to a host of contributing factors such as urbanization and other human land use issues, as well as climate variability and resource availability. Nevertheless, many studies show how restoring keystone species creates powerful food web changes, which include an increase in biodiversity. Justina Ray, director of the Wildlife Conservation Society Canada, recommends using a broad variety of conservation tools in addition to large predators to maintain or restore biodiversity—for example, improving habitat, creating reserves, and adjusting human hunting practices to attempt to stand in for predation by animals such as wolves in places where it is not practical to restore large predators.[31]

Widespread habitat loss and extirpation of keystone predators attendant to expansion of the human enterprise worldwide provide examples of how the premises of the theory of island biogeography, working in concert with Paine's keystone concept, can inform future conservation strategies. The theory of island biogeography asserts that larger habitat patches will harbor more species; Paine's keystone concept is based on how a top predator can by its presence change the density and behavior of its prey in a manner that creates habitat for many other species. Using these two key ecological theories together would mean managing our resources to create larger reserves *and* to restore keystone species in these reserves. Each strategy alone will increase biodiversity; together they are likely to be even more effective. It makes sense to use keystones, trophic cascades science, and whatever tools we have on hand to create more resilient, adaptable populations and ecosystems to deal with inevitable global change. Becoming aware of past and impending losses is the first step; using the science we have within our reach to restore ecosystem resilience and repair the web of life represents the next step.

CHAPTER SEVEN

Creating Landscapes of Hope: Trophic Cascades and Ecological Restoration

My father died of Alzheimer's disease not long ago. He came from a ranching background, although he kept that to himself and I didn't learn about it for many years. His father, my grandfather, had owned a ranch in northern Mexico so enormous it extended from horizon to horizon. It was located in the Sierra Madre Occidental in the state of Chihuahua, near the Sonoran border, the same area Aldo Leopold explored in the 1930s and found so full of wildness. My grandfather was a hard-drinking womanizer who won the ranch in a poker game in the late 1920s. He held on to it for nearly two decades, through my father's childhood, and then lost it the same way he'd won it—in another poker game. This event so profoundly traumatized my father that he never talked about it. I first learned about it when I was in my late twenties and my husband and I decided we'd had enough of city life and bought a ranch of our own. After that my father gradually shared snippets about life on my grandfather's ranch. He had attended boarding school because the ranch was so remote there were no schools nearby. When school ended for the year, he would take a train to the city closest to the ranch, Camargo, Chihuahua. My grandfather would be waiting at the station with my father's horse and they'd ride to the ranch, which took two days. Once there, my father would spend the summer helping with the cattle and hunting to provide wild game for the cowboys to eat.

Alzheimer's is a strange disease. My father was in his late seventies when he was diagnosed. As his mind unraveled he predictably lost his ability to remember the details of daily life: what he'd had for breakfast, where he'd put the television remote control. But ironically his ability to recall events from the deep past sharpened to the point that it became uncanny. With his mind liberated from the minutiae of daily life, old memories surfaced like quicksilver flashes from a mental abyss. As his disease progressed, unfettered by inhibition, he began to tell more stories, among them ones about wolves.

His main job on the ranch had been to shoot varmints on sight. This meant anything with sharp teeth and claws: wolves, grizzly bears, cougars, and of course coyotes. He would spend the day riding to make sure the calves were all right. An avid reader, he carried books in his saddlebag, especially ones by his favorite author, Ernest Thompson Seton. He'd watch the cows and read Seton's wild animal tales to pass the time. At one point my father confessed to me something he'd never told anyone. Sometimes he had seen wolves. My grandfather had impressed on him that above all he was to kill wolves, but my father could never bring himself to do it. To be sure, he said the wolves never molested the cows—at least not on his watch. They always seemed to be moving through the herd on their way to someplace else. They didn't behave aggressively toward him. And there was something about them, in their eyes, in the way they carried themselves, that compelled him to let them go peacefully. Maybe it was all that romantic Seton nonsense about wildness. But quickly he realized that his direct experiences with wolves did not resemble my grandfather's nor the ranch hands', so he kept them to himself.

I clearly remember the moment I told my father that for my PhD I would be studying how wolves affect ecosystems—including areas of multiple human land uses. His mind was three-quarters gone by then. But he got a big smile on his face and said, "Yes! You do that. I'm glad they're back; they are such amazing creatures. But there is still so much we don't understand about them." Then he went on to tell me more wolf stories.

My father inspired me to deepen my understanding of ecological relationships. Time spent afield has taught me that keystone predators, such as wolves, touch all species in an ecosystem. By changing the density and behavior of prey, keystones create habitat for many other species, such as songbirds. In learning about these relationships, which go far beyond the discrete act of a predator killing prey, I have learned that one can use keystones to restore whole ecosystems. Bringing keystones back, because of their far-reaching effects, is one of the simplest ways to improve ecosystem function and increase biodiversity. In this chapter I examine how we can make ecosystems whole again, and provide examples of what I call landscapes of hope.

Integrating Trophic Cascades and Ecological Restoration

What does it mean to restore ecosystems and habitat? And how can trophic cascades science, especially the concept that ecosystems are structured from the top down, with keystone predators playing an apex role, be used to return biodiversity and ecological functions to damaged landscapes? Human activities have disrupted ecological processes by removing natural disturbances such as fire and flooding; by introducing disturbances such as mining, logging, and livestock grazing; and by removing keystone predators, such as carnivorous fish and wolves (*Canis lupus*). All these have been done, and are still being done, to support the growing human enterprise, not always in ways sustainable over the long term. However, one important function of biodiversity, ecosystem services, which include nutrient flow, decomposition, and production of clean water, can be improved by restoring keystone species to landscapes vastly transformed by humans.

In the movie *Field of Dreams*, Kevin Costner's character builds a baseball diamond. Eventually long-dead players show up to play, not all of them visible to humans. In ecological restoration, the field of dreams hypothesis suggests that if you build it, they will come. But restoring ecosystems involves far more than simply restoring physical structure (e.g., returning logs to streams to improve fish habitat), because of the complex dynamics that create and continually modify that structure.[1] Reconsidering this hypothesis means looking deeper and using what we know about ecosystems and keystone species to improve function and resilience. In the past four decades ecologists have acknowledged predators' powerful role in structuring communities and how their removal leads to degraded ecosystems. A degraded ecosystem has reached an alternative state, one usually reversible if given effective physical and biotic manipulation. By restoring keystone predators we can tip many of these systems back to a healthier state with minimal manipulation of physical structure. Kelp forest recovery following the return of sea otters to various parts of their historical range in the North Pacific provides compelling evidence of this. Landscapes of hope such as this provide a brighter future, one with more diversity and resilience than otherwise likely.

Ecological restoration is the process of assisting the recovery of an ecosystem that has been degraded, damaged, or destroyed.[2] However, in any act of restoration it is never possible to return exactly to what once was; one can only move forward. This means recovering a natural range of variation of composition, energy flow, and change, bringing a system back to its historical trajectory. Historical trajectories are only that, since we can't predict the future. We can only work with what we think will optimize adaptability, resilience, and productivity. The past isn't a blueprint for the future, but we can assess these historical trajectories and think about management for future change. This calls for restoring to landscapes as much of their functional diversity as possible, which often means including top predators. Restored systems ideally will be self-sustaining and resilient, exchanging energy with interconnected ecosystems and migratory species. These systems should contain all functional groups (plants, herbivores, predators) and should support reproducing populations of the species necessary for their continued development and resilience.

The science of restoration ecology provides the framework to support these conservation efforts. Areas of food web research most applicable include trophic cascades in lakes and salt marshes, mesopredator release, and keystone reintroduction. Research on these topics has shed light on how reestablishing trophic linkages benefits and strengthens the function and resilience of an entire community.[3]

Restoring Stability and Resilience

Resilient ecosystems have built-in mechanisms that enable them to recover from stress. This means they have sufficient diversity of species and ecological processes that most stresses and disturbances do not result in long-term transformation into entirely different kinds of ecosystems. Examples would be forests that, given sufficient time, return to essentially the same kind of forest following major storms or severe fires. Nonresilient ecosystems, on the other hand, have been pushed far beyond their ability to recover naturally and lack long-term stability. They usually have lost a significant amount of native vegetation and have altered abiotic (nonliving) features, such as topography and soil, which limit their recovery potential, depending on how extensively their species and process diversities have been affected. Ecological restoration calls for removing impediments to natural recovery.[4]

A *reference site* contains a full complement of the species, ecological mechanisms, and interactions to be restored. It can be a physically real site or an ideal compilation on paper. Restoration begins from this guiding image, moving toward clearly defined goals to achieve a healthier, more robust system. This process starts by repairing community structure, monitoring its development, and verifying that linkages between structure and function have been established. It involves acknowledging that the classic view of succession through predictable stages proposed by early ecologist Frederick Clements does not apply universally. It further requires pre- and postassessments to track progress.[5]

Professionals in the field of ecological restoration focus on designing projects to create a well-functioning community of organisms, optimizing both ecosystem services *and* economic value to humans. Not necessarily based on historical conditions, some of these projects are designed to mitigate losses and unfavorable conditions using technology and desirable native species.[6] The Indian Bend Wash Greenbelt in Scottsdale, Arizona, provides an example. Here vegetated pathways and wetlands minimize flood damage from natural stream flow, improve water quality and land values, and create a parklike setting for human recreation.[7] While these sorts of systems may be considered artificial, they do replace ecosystem functions that have been damaged by destructive human land uses. In contrast, systems where the principal change is the recovery of keystone species provide a minimally invasive, highly natural, and cost-effective conservation strategy.

Restoring Lake Mendota with Trophic Cascades

Lake Mendota, possibly the earth's most studied lake, lies in south-central Wisconsin. It is the uppermost, deepest, and largest of the four glacial lakes in the Yahara River chain, near Madison. Condominiums and expensive homes line the lakeshore. The University of Wisconsin and the state capitol occupy the isthmus created by this lake. On a typical summer day, dozens of sailboats dot Mendota's deep blue water. To the average observer the lake looks like a beautiful jewel in an urban environment. University of Wisconsin limnologists Stephen Carpenter and James Kitchell have spent many years studying lakes, however, and they see a lot more. They have long been aware that this lake, overrich in nutrients, represents an off-kilter system. Noxious algal blooms are its biggest problem.[8]

Lakes exchange nutrients and organic matter with land through the downslope movement of water, creating a bottom-up flow of energy that combines with top-down forces (e.g., the effects of carnivorous fish). Adjusting the fish harvest causes energy to cascade through food webs from the top down.[9] However, human land uses such as agriculture and urbanization make nutrients, such as phosphorus, flow into the water. The bottom-up energy thus created negatively affects lakes. Carpenter considers phosphorus the leading cause of lake degradation. Other bottom-up effects, such as climate variability, which makes lakes more hospitable to invasive species, including zebra mussels (*Dreissena polymorpha*) and the Eurasian water milfoil (*Myriophyllum spicatum*), exacerbate damage caused by phosphorus inputs.

Lake eutrophication, due to runoff from agriculture and development, causes a blue-green algae bloom. These algae, also called *cyanobacteria*, are microscopic photosynthetic bacteria that occur naturally near the water's surface. We have seen how an influx of phosphorus stimulates them to multiply quickly and bloom during the warm months of the year. These blooms negatively affect the quality of water, making it more turbid, and present serious threats to the health of humans and other mammals, including allergic reactions, liver toxicity, and even death. Managers can counteract algal blooms by adjusting fish harvest levels and stocking rates.

Biomanipulation is based on the trophic cascades premise that the amount of algae reflects the intensity of grazing by herbivore populations, which are controlled by predators from the top down.[10] University of Wisconsin limnologists and researchers from the Wisconsin Department of Natural Resources (WDNR) initiated a whole-lake biomanipulation project on Lake Mendota in 1986 to see how this would work. In this four-level trophic system it involved stocking the lake with walleye (*Sander vitreus*) and northern pike (*Esox lucius*) fingerlings over a thirteen-year period. These two piscivorous (fish-eating) species at the top, or fourth, trophic level prey on planktivorous (plankton-eating) fishes, such as cisco (*Coregonus artedi* Lesueur) and yellow perch (*Perca flavescens*), at the third level. Cisco and perch feed on *Daphnia*, the most common grazers in lakes, at the second level, and *Daphnia* eat algae at the first level. In something of a domino effect, increasing piscivores reduces planktivores and increases *Daphnia*, thereby reducing blue-green algae. WDNR limnologist Richard Lathrop found that in practice these food web effects cascaded down to lower trophic levels, improving water quality. However, they were not uniform but were complicated by phosphorus inputs, which fed algae, as well as challenges in stocking and maintaining fish populations at desired levels. But in general biomanipulation has been an effective way to restore Lake Mendota to more desired conditions.[11]

This was the first lake to be biomanipulated. One of the leading questions has been whether the results justified the several million dollars in expense and the magnitude of this effort. A cost analysis found that it yielded valuable results, with expenses recovered via ecosystem benefits, such as clean water for the city of Madison and healthy, harvestable fish. Like all lakes, Mendota is a complex system, with many unexpected events occurring and neither top-down nor bottom-up processes prevailing. These effects interact, alternate in importance, and operate within different temporal and spatial scales.[12] Kitchell and Carpenter recommend further research to increase our understanding of these interactions.

Restoring Salt Marshes with Trophic Cascades

Barrier islands are long, narrow bodies of land, found worldwide, that run parallel to the mainland. Sculpted by wave action and ocean currents, these offshore islands help protect the coast from the erosion caused by surf and tidal surges. Sapelo Island is one of the largest and most pristine of Georgia's barrier islands. Formed during the Pleistocene epoch, this island contains the Sapelo Island National Estuarine Research Reserve, a coastal plain estuary that comprises marsh, maritime forest, and beach dune habitats. This important research outpost forms part of the National Science Foundation's Long-Term Ecological Research network.

Salt marshes are among the earth's most ecologically and economically productive ecosystems. They are threatened by increasing exposure to human disturbance, such as runoff from agriculture and introduction of invasive species, and are experiencing a massive die-off worldwide. Drought and herbivory by species such as the marsh periwinkle (*Littoraria irrorata*), when unchecked by predation, present additional threats. University of Florida ecologist Brian Silliman and his colleagues have worked since the late 1990s to investigate how unimpeded periwinkle herbivory on grasses quickly turns lush, green salt marshes into barren mudflats via a truncated food web. They wanted to know how overfishing of marine predators cascades downward to affect the structure and function of salt marsh ecosystems.

Spartina is the dominant grass in the Sapelo Island salt marsh. Until recently scientists thought that bottom-up forces prevailed in this system, with herbivores of little consequence to community structure. However, none of the early studies tested what would happen if grazers were removed. In fact, researchers considered *Spartina* relatively resistant to most forms of herbivory until the 1980s, when they discovered that wild horses, nutrias (*Myocaster coypus*), and insects graze on this grass.[13]

The food web Silliman and Mark Bertness studied consists of marine predators at the top level: blue crabs (*Callinectes sapidus*), mud crabs (*Scylla serrata*), and terrapins (*Malaclemys* spp.). Marsh periwinkles make up the second trophic level, followed by *Spartina* on the first level.[14] Silliman and Bertness manipulated predator and herbivore densities using cages that excluded crabs and terrapins. The cages contained three different snail densities: no snails, moderate snail density, and high snail density. When the researchers excluded predators, periwinkles severely overbrowsed *Spartina*; in areas outside cages, where predators interacted with periwinkles, *Spartina* grew lush and tall. Silliman and Bertness further correlated intensity of herbivory in the absence of predators to snail density, finding marsh grasses more heavily grazed in cages with high snail densities than in cages with medium snail densities. This experiment suggested that marine predators regulate grazer abundance, which regulates grass growth. And it demonstrated that truncated trophic cascades may be a key factor in salt marsh die-off.

Beyond Yellowstone: Zion Canyon

Few better examples exist today of a reference landscape than Yellowstone National Park, a previously ecologically damaged system that has been partially restored by simply adding one species—the gray wolf. Since 1996 Robert Beschta has been looking at hydrologic degradation in Yellowstone caused by intense browsing by elk (*Cervus elaphus*). In 2002 he began collaborating with William Ripple to study these effects in riparian communities. In relation to the ecological cascades triggered by the wolves' return, some of Ripple and Beschta's most relevant work examines how these principles can be applied to ecosystem restoration. "On a practical basis, trophic cascades can help us understand how we can use these processes to restore ecosystems," Ripple explained. "Can systems simply flip back to being fully or partially functional? What happens when you lose some of the species or some of the processes; can they be restored?"[15]

Trees, shrubs, and grasses that grow on stream banks are essential for riparian health. According to Beschta, "when a river floods, the water in the main channel is flowing very fast. But then it overflows its banks and goes out onto the floodplain, and the vegetation there slows the water and enables it to deposit silt, which enriches the system. When you have an overbrowsed system without large predators, you get a river that causes massive erosion when it floods." Beschta suggests that loss of beavers (*Castor canadensis*) in the West has had a profound impact on riparian ecology, but even more significant has been the loss of wolves. He points out damaged ecosystems where wolves have been extirpated and there are too many cattle. The combination of livestock and elk herbivory in areas of multiple human land use makes matters worse because elk really are not so different from livestock in terms of the damage they can inflict on a range if unchecked by predators. "Herbivory is perhaps the most underrated force in ecosystems across the world," explained Beschta. "If people really knew how much plant communities have changed following the loss of major predators, they would be shocked."[16]

Ripple and Beschta have done most of their research in protected areas, working in six national parks in two countries. They find parks easier to study because they contain fewer confounding factors, such as human land use and livestock herbivory. But Beschta points out that the same ecological principles apply in all systems. One of the most important is how loss of top predators indirectly causes erosion and ecosystem simplification.[17] By 2005 they began to look farther afield for the patterns they'd found in Yellowstone and were drawn to the Southwest by Aldo Leopold's work there on ungulate irruptions.

WHEN I was working on my master's degree in environmental studies, I attended Prescott College in Arizona. On my way there and back from my Montana home, I would take side trips into Utah's red rock backcountry to hike and photograph the stunning sandstone formations. Sometimes I would stop in Zion National Park and visit the more remote side canyons. A walk I took one spring stands out from the others.

Red sandstone rose all around me as I entered one of the narrow river-cut canyons in the park. Under the songs of blue grosbeaks, plumbeous vireos, and yellow warblers ran the sound of water flowing over stone. Sand underfoot held the tracks of deer and ringtails as well as the footprints of a young child running alongside her parent. I could almost hear that child's laughter bouncing off the curving canyon walls. Cottonwood leaves shimmered a translucent tender green in the luminous May light. Beyond them the canyon's russet walls rose in a sharp V. A clump of *Datura* grew from a hanging garden a hundred feet above the canyon floor, jutting out from a sheer rock face. A potent hallucinogenic also known as jimsonweed, *Datura* has white, trumpetlike flowers that open at dusk, close at dawn, and are pollinated by moths. Tricks of the light, ancient echoes, intimations of immortality. I wondered how many ages it had taken for the river to carve this canyon. I walked on, pausing periodically to write and sketch in my nature journal, trying to capture the essence of this place. I left the narrow trail and sat by the river, experiencing its flow for what felt like long moments. When I looked at my watch I discovered that only ten minutes had passed. These canyons can do that to you—change the way you perceive time, make modern constructs lose their importance in the face of wonder.

Back on the trail I found a perfect cougar track pressed into the sand. I'd never seen one more clear, so I lifted my camera to photograph it. But I never took that picture because I found a second one next to it, on top of one of the tracks I'd made on my way into the canyon. The cougar wasn't stalking me; it was walking in the opposite direction that I had been. However, it had come within fifty feet of where I'd sat on the riverbank. All at once that track put me in touch with wildness so deep that nothing else mattered. I walked out of that canyon aware of every sound, every movement of the light. I never did see the cougar. Later I read in a guidebook that they sometimes follow hikers through these slot canyons, more out of curiosity than anything. When I closed my eyes to sleep that night, the image of that track, its three-lobed metatarsal pad pressed cleanly into red sand, lingered in my mind.

CREATED IN 1919, Zion is Utah's oldest and most popular national park. As in many parks, humans concentrate their activity in one area—in this case Zion Canyon. Ripple and Beschta came here in search of another trophic cascade, and they found one. It was like the Lamar Valley again, only this time the potential keystone species was the cougar (*Puma concolor*) and the herbivore was the mule deer (*Odocoileus hemionus*). Different system, same dance.

Cut long ago by the North Fork of the Virgin River, Zion Canyon runs fifteen miles long, as much as a half mile deep, and several miles wide. It harbors a remarkable diversity of species because it lies at the juncture of the Colorado Plateau, Great Basin, and Mojave Desert ecoregions. Since the park's inception this canyon has been a hub of human activity. Ripple and Beschta compared the ecology of the North Fork of the Virgin River to that of another stream in the park, neighboring North Creek, which runs seven miles through a narrow, roadless canyon seldom visited by humans. North Creek is one of many that flow into Zion's backcountry, sculpting deep canyons bounded by sheer Navajo sandstone cliffs. Ripple and Beschta hypothesized that cougars would be rare in Zion Canyon, because of high human presence, and common in North Creek, which has low human presence. They did not use radio-collar data but based their analysis on a correlation between historical and current wildlife information and the plant, invertebrate, and amphibian data they gathered in their study sites. The contrast they made between the two areas was especially relevant because this study, conducted in 2005, represented the first time anyone had identified the cougar as a keystone.

The differences were astounding: along North Creek they found 47 times as many cottonwood (*Populus* spp.) trees and 5 times as many butterflies. North Creek held 200 times more toads and frogs. It harbored cardinal flowers (*Lobelia cardinalis* L.) and asters (*Aster* spp.) by the hundreds, while Zion Canyon plots had none. North Creek cottonwoods showed continuous recruitment, while those in Zion Canyon featured the now familiar pattern of missing age classes. Ripple and Beschta correlated this difference to the cougar's possible decline in Zion Canyon, which allowed deer numbers to soar, leading to loss of cottonwoods, eroding stream banks, and a sharp decline in biodiversity.[18]

There is some uncertainty in science about why cougars don't function as keystones in most ecosystems (e.g., in Yellowstone) but do in some (Zion). Social habits and historical predator presence may have something to do with this. Cougars are solitary stealth hunters that don't form social groups, while wolves form packs and hunt in groups. This means that where they occur together, wolf presence has greater impact than cougar presence on top-down trophic interactions. Multiple species of predators (also called predator guilds) historically present may also influence these interactions. In Yosemite National Park, Ripple and Beschta found another cascade involving cougars, mule deer, and black oaks (*Quercus velutina*).[19] There the cougar possibly takes on the keystone role because there may never have been wolves in that system. But where wolves have historically been present, as in Yellowstone, cougars have not been found to drive a trophic cascade. However, wolves were historically present in Zion National Park, yet researchers documented a cougar-driven trophic cascade there.

Cougars have an undeniably strong effect on ecosystems, but science has a fledgling understanding of exactly what that role may be. Cougar cascades represent a frontier in trophic cascades science—an area where much more research is needed to determine these large predators' role in shaping food webs. Working with data from radio-collared cougars will help shed more light on these interactions. Research is under way to investigate these relationships in greater depth.[20]

Trophic cascades principles can be applied outside national parks to restore working landscapes suffering from excessive herbivory. A key concept used by Ripple and Beschta and others is particularly useful for ecological restoration: *refugia*—places where palatable species are protected from ungulate browsing.[21] Refugia can help managers and conservationists understand the role of ungulate herbivory and can be used as reference sites. Fenced exclosures provide the ultimate refuge but have some limitations. Because of their typically small size (e.g., a quarter of an acre or less) they may contain atypical plant communities and can't fully represent undisturbed preexisting conditions. Refugia may be most useful where browsing is done primarily by domestic ungulates; where wild ungulates are present, with large predators missing; and to evaluate variation in predation risk where predators are present.

Ecological Restoration on Working Landscapes

While sound science and cultural history should provide the foundation for any ecological restoration project, decisions regarding where, when, and how to conduct restoration will always be based on current social values.[22] With most restoration projects, multiple stakeholders from various backgrounds have an interest in the project and its long-term implications. These stakeholders may include surrounding landowners, recreational users, government agencies with jurisdiction over the property, and special interest groups.

North American ranches offer some of the best examples of how one can gather a diverse community of people to bring about ecological restoration. For example, predator-friendly ranches managed for multiple compatible uses create opportunities to study how repatriated wolves might affect ecological relationships in these working landscapes. Here I present examples of three ranches being managed to create and preserve intact ecosystems: a ranch owned by a conservation organization, a family ranch, and a corporate ranch.

Theodore Roosevelt Memorial Ranch

The Theodore Roosevelt Memorial Ranch (TRMR) lies southeast of Glacier National Park and the Blackfeet Indian Reservation in Montana, where short-grass prairie meets the sharp limestone upthrust of the Rocky Mountain Front. The Boone and Crockett Club owns and operates this ranch as a place-based conservation education center and a research station affiliated with the University of Montana's Wildlife Biology Program. The TRMR is a demonstration ranch, an example of how conservation and ranching can work together—quite a challenge, considering that grizzly bears (*Ursus arctos*) and wolves live in this wild landscape, along with cougars, a large migratory elk herd, a large population of migratory mule deer, and residential white-tailed deer (*Odocoileus virginianus*).

I headed west on a gravel road from the small ranching town of Dupuyer toward the ranch, which lies deep within the foothills that crest and break against the Rocky Mountain Front. Late autumn sunlight brought silvery aspens and inky conifers into sharp relief against gentle mounds of prairie. Dark forms of grazing Angus cattle dotted the hills, part of the 200 cow-calf pairs stocked on this ranch. Tidy haystacks put up for winter lay behind an old ranch house, now used as a bunkhouse. This land's pastoral beauty belied its hard history. Homesteaders came in the late 1880s, looking for paradise. What they found instead was an unforgiving landscape as brutal as it was lovely. They faced harsh winters, drought, and disease; few lasted long.

TRMR manager John Rappold is a fourth-generation descendant of the stalwart German homesteaders who stuck it out. Sturdily built, with clear eyes the color of the famed Montana sky, fair hair, and ruddy skin weathered by the sun, he reflects this landscape in his physical appearance. His family's ranch lies just north of the TRMR. John's management style combines bred-in-the-bone wisdom with the state-of-the-art range management and ecological principles he learned in college. This approach is yielding a place where cattle coexist with large carnivores.

The TRMR was formerly the Triple Divide Ranch, purchased by the Boone and Crockett Club in 1986. It totals 6,000 acres abutting the Lewis and Clark National Forest along the Rocky Mountain Front, with another 4,500 acres of grazing allotment. The club's policy calls for shared, sustainable, and ethical resource use. Previous ranch owners had grazed this land very hard. The first manager, Robert Peebles, a third-generation descendant of one of the families that homesteaded this area, started the TRMR land on the road to ecological recovery. Since 2000 Rappold has been managing the ranch to increase cattle fitness, improve range conditions, and promote livestock coexistence with mule deer, elk, and a full complement of large predators. And it breaks even financially. As he explains it,

> We try to keep the ranch and everything as healthy as we can. People stop by and say, "Well, yeah, but if this was your own ranch, you'd run more cattle, wouldn't you?" And I say, "No, I wouldn't, not unless there was some way to do it sustainably." I want to see the aspens healthy, I want to see the brush in the creek healthy, I want to see healthy riparian areas. I want to see quite a bit of grass left in winter, to hold snow and ease erosion in spring runoff time. And I like that there are bears and wolves here. There are areas on the ranch that I'd like to look nicer, and I'm making those adjustments. It is the Boone and Crockett Club's ranch, but it pretty much operates as if it were a regular family ranch up here.[23]

Predator presence is moderate, with one wolf pack ranging in the area and several grizzly bears very actively using the ranch in spring and summer. According to Rappold, bears and wolves have their place on the TRMR, but management of these species needs to be in place here too. This could entail nonlethal as well as lethal methods of animal control, although predators have not yet presented a problem here. The biggest challenges Rappold faces are not predators but finances. These days it's getting tougher to run cattle and make ends meet. Stories about ranching and hardship are common in the West. While the Rappold family held fast to their land, others were unable to do so.

The American game conservation movement originated in the late nineteenth century when hunters, alarmed about declining wildlife populations, began demanding stronger fish and game laws. But it wasn't until 1883, when Theodore Roosevelt visited North Dakota searching for a prize buffalo trophy and realized that the vast migratory herds, once icons of the American West, were gone, that this movement gathered momentum. In 1887 he cofounded the Boone and Crockett Club to conserve as much of our American wild heritage and hunting values as possible. The first private organization to address conservation issues nationally, the club convened prominent sportsmen, industrialists, politicians, scientists, and writers to create a vision of sustainable fish and wildlife resources.

From the beginning the Boone and Crockett Club has been about balancing human and wildlife needs. Members such as George Bird Grinnell, Gifford Pinchot, John Lacey, and later Aldo Leopold laid the foundation for the club's conservation program, which initially involved eliminating large predators. Gradually, as a result of Leopold's influence and the work of contemporary scientists, this vision began to include wolves.

Since the mid-1980s club professional members such as Jack Ward Thomas, former University of Montana Boone and Crockett Professor Hal Salwasser, University of Montana Wildlife Biology Program director Dan Pletscher, and Paul Krausman, who holds the club's endowed chair at that university, have used the TRMR to advance the organization's conservation objectives. In the 1980s club leaders created a vision for the ranch and professorship, which they matched with the money needed to fulfill it. They operated from the premise that public lands alone could not sustain the full array of large mammals in the West. Major allies include The Nature Conservancy; Montana Fish, Wildlife, and Parks; Plum Creek; and Qwest Communications, which contributed to the purchase of the TRMR. Salwasser guided the baseline research on the ranch's ecosystems and cultural history as well as wildlife policy research about how private lands contribute to conservation goals. He also created the TRMR conservation education program. According to Salwasser, the ranch still has a way to go on riparian restoration, and fire suppression presents another issue. The Rocky Mountain Front had more frequent fire centuries ago, set by the Blackfeet to improve hunting and habitat for wildlife.

Krausman is developing a monitoring strategy using remote sensing data to examine and predict habitat use by native and domestic ungulates. He is working with Rappold to manipulate livestock use as much as is economically feasible. Krausman's vision involves demonstrating what would be needed to operate a successful cow-calf operation while enhancing and maintaining quality habitat for wildlife and documenting the trade-offs. This means identifying the tipping point where livestock has a negative influence on wild ungulates by overbrowsing and overgrazing riparian areas. He has initiated a mule deer study to examine a host of concepts, including migration, herd condition, and habitat, and is documenting current predator use of the TRMR.

Environmental education programs on the ranch, directed by Lisa Flowers, enable schoolchildren, their teachers, and adults to experience what stewardship can look like in a system that, as Aldo Leopold put it, contains all the pieces. Each year the Elmer E. Rasmuson Wildlife Conservation Center for Education and Research, a state-of-the-art facility, welcomes 300 teachers, 2,000 students, and other civic and conservation organizations. Krausman's ecological monitoring program ties in with the education program to enable teachers and students to contribute as citizen scientists to conservation. These programs provide an inspiring example of how one can integrate agriculture, wildlife science, and the land ethic to sustain healthy ecosystems.

The Russell Ranch

The Russell Ranch perches on a promontory on Waterton Lakes National Park's northern boundary, in Alberta, Canada. Like the TRMR, this Rocky Mountain Front ranch harbors a full suite of large predators, including wolves and grizzly bears. It abuts critical winter range for the large elk herd that ranges in southwestern Alberta, just south of Shell's Waterton Gas Complex, established in 1957. The East Slope of the Rockies is changing rapidly as a result of oil and gas development. Homesteaded in 1905 by Bert Riggall, this ranch and most of the neighboring lands remain intact because locals have banded to keep out subdivisions and commercial development.

The Riggall-Russell family began as guides and outfitters in Waterton, eventually becoming environmental activists, wildlife filmmakers, bear and caribou experts, and writers. Bert's son-in-law Andy, born in 1915, was widely known as a mountain man, cowboy, and writer. One of the most engaging storytellers and passionate conservationists in Canadian history, he dropped out of high school during the Great Depression to make money by trapping wildlife and training horses. After years of actively killing predators, as was customary in the middle of the past century, he became among the first to promote conservation of large carnivores in Canada. He also fought to stop development of the East Slope of the Rockies and to preserve wildness and the ranching way of life, writing fourteen books in the process. Andy died in 2005. His family now consists of his four sons and a daughter and their spouses and children, several of whom still live on the family ranch.

Andy's sons have followed in his footsteps. John Russell is a biologist who has written a book on caribou. He trained as a zoologist and has studied caribou, mostly in the subarctic, for more than twenty-five years, and he did one of the earliest bear DNA studies in Alberta. Currently he serves as a councillor for the local government to the north of Waterton Park. As a politician his main responsibility involves making land use decisions. Charlie Russell has spent much of his life working with grizzly bears in Canada, Alaska, and Kamchatka, Russia. He focuses on conservation and human communication and peaceful coexistence with this species. He has written and produced several books and films about grizzlies. The Russells have worked intensively with The Nature Conservancy and other organizations to protect lands along the East Slope.

The main ranch house, the Hawk's Nest, sits on a mountaintop and offers a 360-degree view of this landscape's legendary wild beauty. Over dinner there one night, John explained his family's feelings about conserving this land: "Our vision here is to keep this land the way it was in a natural state, do the best we can with what we have. Our objective is to maintain this more or less as wilderness, but with cows as the main herbivore."[24] This represents a huge shift since Bert's time; Bert saw his role, along with everybody else's, as to "unwild the wilderness" by removing the top predators. As we talked that night we could hear a female cougar, lying in the shrubs a stone's throw from the Hawk's Nest, repeatedly yowling for her mate. The Russell family's calm reaction to her passionate cries emphasized how far they have come in terms of welcoming predators since their grandfather's era.

The Russell family leases the grazing for eighty-two cow-calf pairs for three months (July through September) on two sections (1,280 acres) of land in what is mostly aspen (*Populus tremuloides*) forest. Range conditions are generally good, with riparian areas fenced off to protect them from overbrowsing and stream bank erosion. The Russells have not experienced much depredation on their ranch over the years. Charlie thinks it has to do with the good relations they maintain with predators, which begin by their not automatically taking an adversarial stance toward them. Grizzlies often prey on livestock when they waken from hibernation in the spring, targeting calves, but the stock on the Russells' ranch has not been troubled much by them. John believes this may be a result of animal husbandry practices that involve handling the calves more gently at branding time than do most ranchers. The logic behind this is that injured, vulnerable calves are more likely to attract predators. The Russell Ranch lies in some of the best bear habitat in Alberta. The Russells let the bears forage on their land and don't shoot them or fear them but treat them with healthy respect.

Wolves are present on the ranch, denning nearby. However, the wolf population here is kept moderately low because in southwestern Alberta it is legal to shoot and trap wolves. While some of their neighbors kill wolves, the Russells strive for peaceful coexistence. They see this as a matter of maintaining clear and respectful communication with wildlife and a willingness to adjust husbandry to improve these relations. For example, they quickly dispose of the carcasses of livestock that have died of natural causes, to avoid attracting predators. Who knows what Bert would make of all this if he were alive today, but he would be pleased to see the ranch he created intact and in good condition, with healthy livestock and game.

The High Lonesome Ranch

The High Lonesome Ranch occupies a blank spot on the map in the northwestern corner of Colorado, on the West Slope of the Southern Rockies. It encompasses several valleys and mountain ranges, totaling approximately 300 square miles of deeded, permitted lands. Ecological communities include aspen parkland, sage flats, scrub oaks, wetlands, and grasslands, providing diverse game habitat. Paul Vahldiek Jr. and his wife, Lissa, began acquiring this land in 1994, piecing it together from several ranches, many of them owned by descendants of the valley's early homesteaders. With conservation their primary objective, they set out to manage this land for biodiversity—a mixed-use vision that includes cattle, mineral development, and large predators. The High Lonesome Ranch offers a model for sustainability that provides stewardship of a large-scale, intact western landscape. To fulfill this vision, Vahldiek, his partners in the ranch, and ranch managers are restoring species and damaged habitat and ensuring long-term conservation of critical open space while engaging in enterprises that support the ranch. Indeed, Vahldiek's definition of conservation involves humans. Over supper one night with the Russell family at their Alberta ranch, he commented, "We are in the environment; we are not leaving. So how do we wake up and still see elk in the morning, and eagles, butterflies, and leopard frogs? Y'all are not giving up your ranch; I am not giving up mine. So how do we coexist and do it at a reasonable, sustainable level?"[25]

In the West, open space is essential to maintain large predators and the intact ecosystems necessary for top-down trophic cascades to function effectively. According to Vahldiek, who is a tax attorney, what's really leading to the breakup of many great ranches is the United States' arcane death tax. He explained: "Let's say that I was a father who owned a ranch. The ranch was in my name, valued at over two million dollars, and I wanted to leave it to my two sons. Upon my death it would be taxed at the rate of 45 percent. So how do you pay that, besides breaking up the ranch and selling it?"[26] The Nature Conservancy offers conservation easements, which provide a way for heirs to avoid paying inheritance taxes. Vahldiek and his partners have opted instead to create a corporation to manage and place an easement on the High Lonesome to ensure that the land will be conserved in perpetuity. But maintaining large open spaces by preserving ranches is only the beginning of the solution. The next step involves managing them in ways that maintain the ecological fitness of private lands.

Northwestern Colorado was historically renowned for mule deer and elk hunting. As in other places, in the early twentieth century both of these species suffered from overharvesting but have since rebounded. Vahldiek, his partners, and ranch operations manager Doug Dean, who has degrees in rangeland science and business, have continued to restore the game population using progressive, ecologically sound range principles. They have been replanting grasses to create food plots for cattle and wildlife and to provide barriers against spring runoff. Dean reduced the number of cattle from a historical stocking rate of approximately 1,250 cow-calf pairs to 350, moving them frequently to emulate grazing patterns in places like the Serengeti. To support this restoration Vahldiek and his associates supplement ranch income by providing guest services, which include big game hunting, wing shooting, fly fishing, equestrian sports, environmental education, and ecotourism, such as birding and wildlife tracking.

Dean took me on an ecological tour of the ranch. At an overlook I stood on a rocky point and gazed out at vast ranchlands in the foreground, Bureau of Land Management grazing allotments in the middle ground, and pockets of gas development in between. The land unfurling at my feet held questions, challenges, and some possible answers. It's a landscape that has long had little room for wolves and other carnivores, but it holds the promise of change. Finding solutions means taking a close look at how natural gas and oil shale development may affect wildlife.

Like other ecosystems that contain elk, aspen, and wolves, the story is writ on the landscape. Following wolf extirpation from this area in 1945, cattle numbers increased, creating a downward ecological spiral that left this landscape depleted. Aspen recruitment into adult trees ceased in many places during the last half of the twentieth century, with little new growth evident in the stands on the upper ranch. Through careful management the High Lonesome now harbors approximately 400 elk, an equally robust but sustainable mule deer population, and a high cougar population. Cougars here do not function as keystone predators, as evidenced by their high abundance and the overbrowsed condition of the aspen, even in areas where cattle are excluded. Vahldiek and his managers move livestock frequently to avoid overlapping them with elk on winter range in the valley bottoms. Nevertheless, aspen recruitment continues to be limited to occasional pockets, often in refugia on steep terrain inaccessible to elk.

I visited the northern half of the ranch with Dean. Driving seventy-five miles, much of it on High Lonesome property, we made our way through a broad valley and up a series of sharp switchbacks to the top of Kimball Mountain. There we passed through a ranch gate onto a high mesa punctuated by vast aspen parklands. Stand after stand consisted of aging aspens and showed no recruitment. This is a multicausal situation, attributable to disease, drought, and herbivory. As I walked among these aspens I noted how all the young ones had been browsed repeatedly, to the point of taking on a bonsai form, with many having been killed by herbivory. This place provides both summer range for cattle and a calving ground for elk; perhaps this combination in the absence of wolves is more than the aspens can sustain. I found fresh elk sign everywhere—and fresh browse on the aspens.

But, like other landscapes near the Greater Yellowstone Ecosystem, this landscape is changing. Wolves are actively recolonizing this part of Colorado, dispersing south from the northern Rockies. In the winter of 2006 a wolf was spotted twenty miles from the ranch's northern boundary, and in 2008 a ranch manager observed a wolf on Kimball Mountain. In February 2009 a yearling female wolf from the Greater Yellowstone Ecosystem, who wore an Argos radio-collar, traveled 1,000 miles, to northwestern Colorado, in search of a mate. She was found dead one month later not far from the ranch. Nevertheless, more wolves are on the way. It is only a matter of time before wolves become established here, and when they do, High Lonesome will accept them. In the meantime scientists are mapping range conditions and wildlife distribution on the ranch and evaluating the fitness of mule deer and elk herds. Riparian restoration begun by the Academy of Natural Sciences in Philadelphia includes an inventory of indicator species, such as amphibians, to create baseline biodiversity data for this ranch. Additional riparian restoration may be undertaken by the High Lonesome Ranch, Trout Unlimited, the Colorado Division of Wildlife, and Chevron.

I am a part of the ecological restoration efforts on this ranch. Working within a trophic cascades framework, I am measuring aspen stand structure and biodiversity, using songbirds as biodiversity indicators. Additionally, I am using noninvasive methods to measure wildlife distribution and density across this landscape in order to determine the effects of wolf and cougar predation on elk and mule deer. Working with top wildlife trackers, I am putting in track transects and also conducting focal animal observations to measure elk vigilance behavior. Strategically placed remote cameras are providing compelling data in the form of images of the wildlife using this landscape. Other projects will include multipredator interactions involving cougars, bears, and wolves. Michael Soulé and Wildlands Network are helping guide this ecological restoration, which may include bison (*Bison bison*) reintroduction. Bison roamed Colorado widely from the late Pleistocene epoch until the 1880s, when they were extirpated via overhunting. Accordingly, a bison reintroduction on the High Lonesome Ranch would bring back to this system a native large herbivore whose presence has been found to improve grassland communities and riparian health in other places.[27]

As Dean and I drove down Kimball Mountain at dusk, the land felt like wolf country already. I could almost see a line of wolf tracks illuminated in the dust by the setting sun. Indeed, given their predilection for dispersing onto new terrain, soon there may be wolf pups in these mountains.

Trophic Cascades and Multiple Human Land Uses: The Southwest Alberta Montane Research Program

In addition to predator-friendly ranches such as those discussed earlier in this chapter, ecosystem-scale interdisciplinary research on multiple-use, mixed-ownership landscapes is contributing to conservation. The Southwest Alberta Montane Research Program was initiated in 2006 to investigate how multiple human land uses affect elk behavior and habitat. This six-year project spans the US-Canada border and runs along the Rocky Mountain Front. It brings together several government agencies, three universities, and several community conservation groups. Shell Canada, Alberta Sustainable Resource Development, and Parks Canada provided initial funding, which was supplemented by grants from various conservation and government organizations. Taking a community ecology approach to studying elk in this high-tech project involves looking at the web of life, including grizzly bears, wolves, cattle, and humans and their effects on elk.

The Southwest Alberta Montane Research Program was the brainchild of Carita Bergman, a wildlife biologist with Alberta Sustainable Resource Development charged with managing populations of ungulates and large carnivores in southwestern Alberta, and Shell Canada's former principal ecologist, Roger Creasey, who grew up on the Rocky Mountain Front and saw it change tremendously over the years—from privately owned ranches that averaged one square mile in size to housing projects with hundreds of ranchettes on quarter-acre lots, mixed with oil and gas development. Concerned about the effects of habitat fragmentation and human activity on wildlife, Bergman and Creasey designed this project to help create long-term sustainability and a practical conservation agenda. With these objectives in mind they carefully selected a diverse group of scientists, each of whom would provide part of the solution to the problem of sustainable resource management. Primary investigators include renowned elk biologist Mark Boyce and Marco Musiani, an expert on wolf ecology and livestock depredation. Each of the eleven graduate students engaged in this effort studies a different aspect of elk and predator ecology—from road access and elk movements to genetics. This has involved putting GPS collars on 100 elk, several wolf packs, grizzly bears, and cougars. My research focuses on how wolf predation influences elk behavior and habitat via trophic cascades. Additionally, I am working with others to compare elk and wolf interactions in protected areas, such as national parks, to elk and wolf dynamics in areas of multiple human land uses outside parks.

Eventually our findings will be used by government agencies to restore and protect elk habitat and to help maintain as intact an ecosystem as possible in an area of growing human land use. Creasey's lifelong commitment to conserving this landscape has shaped this research program from the start. When I asked him about his guiding vision and desired outcome, he said:

> I expect this collaboration to produce results that help Shell, other companies, and other public government entities plan better. In the past I've seen chaos in the hills. And because I study cumulative effects, I know that everything is related. So I've come to realize that there is a big picture and we won't get there unless we plan along the way. This project will provide better information to help others do the planning. And part of this has to do with making sure that the company I work for and represent is responsible for their impacts and does better. So losing the resident herd on Prairie Bluff or the elk in the Carbondale—no, not on my watch![28]

AS THE human population burgeons and ecosystems become increasingly transformed by human actions, ecological restoration will become more and more important. Trophic cascades principles provide a sound foundation for many restoration projects. While it's not possible to restore keystones to all systems, with enough ingenuity and determination we may find that many places are suitable for restoration based on trophic cascades. This approach can sometimes offer the most effective path to returning resilience and vitality to landscapes.

Where it's not possible to restore keystone species, our best option is to take the lessons we've learned from trophic cascades science and apply them to management of large herbivores. The state of Pennsylvania provides an example of a place where even though it may be impractical to restore wolves, it may be possible to apply trophic cascades principles to manage deer in order to mitigate the damage they are doing to plant communities. This may mean revisiting some of Aldo Leopold's ideas about game management and considering things such as a year-round hunt, to replicate the effects of predators. Radical in his era, these ideas continue to be radical today. But the damage caused by too many deer hasn't changed much in areas lacking keystone predators since his time, and indeed it may have gotten worse. Applying trophic cascades science to manage ecosystems can provide a promising way to solve these problems.

CHAPTER EIGHT

Finding Common Ground: Trophic Cascades and Ecosystem Management

Imagine a world populated by creatures sharp of tooth and claw—wolves, grizzlies, and cougars. A world with linkages between the protected areas that harbor these predators, enabling them to move about relatively freely, reproduce, and live well on abundant prey. A world without excessive deer or other ungulates, one with multistoried forests and nutritionally rich, diverse grasslands. A world where we can heal the damage caused by the extinction spiral that began in the Pleistocene epoch and continues today because of human overexploitation of resources. Imagine a world filled with rich and diverse landscapes that will allow most, if not all, native species to adapt and persist amid changes driven by climate and human actions. Such is the rewilding vision created by the group of conservation biologists and wilderness advocates who founded the Wildlands Project in 1991. Their collective objective was to restore large carnivores to North America. Among them were Dave Foreman and Michael Soulé. Foreman coined the term *rewilding* for this group's idea for restoring wildness and wilderness on a continental scale.[1] Over the next eighteen years, under the guidance of Soulé and others, the Wildlands Project continued to think big in terms of time and space to reconnect, restore, and rewild ecosystems on both public and private lands. Widely regarded as a charismatic conservation leader and activist, in 2003 Foreman founded The Rewilding Institute, a think tank advancing ideas for continental conservation. Both organizations have led this progressive conservation strategy.

Rewilding is based on Aldo Leopold's admonition that the first rule of intelligent tinkering is to save all the pieces. In practice, twentieth-century resource use policies often meant exploitation without awareness of or care about consequences. Science had an incomplete understanding of these consequences, and the few voices that expressed concern, such as Gifford Pinchot and Leopold, were not always heard. Accordingly, today we are faced with the task of "putting nature back together."[2] The Wildlands Project proposed to take on this task by reforming management practices to restore top predators to places where they had been extirpated and by creating the habitat, large reserves, and human tolerance to sustain these species.

In the 1990s ideas about corridors and linkages and top-down regulation by large carnivores were highly controversial, both scientifically and politically. In supporting rewilding during the early years Soulé, John Terborgh, James Estes, and Joel Berger had the courage to voice what some scientists know but usually don't say. They were not the first to do so. Leopold spoke out about the ecological value of predators during an era when these ideas amounted to heresy. Today the scientists participating in rewilding are making statements grounded in rigorous science about far-reaching conservation planning that encompasses all of North America.

Large predators are restricted to a fraction of their former North American range, with many landscapes ripe for rewilding. Widespread issues created by predator removal include superabundant ungulates, biodiversity loss, and development of plant and animal communities that differ markedly from what would occur naturally. At the core of rewilding lies the premise that ecosystems are partly driven by top-down forces, via trophic cascades, and that restoring keystone predators on an ecosystem scale can restore biodiversity via the many indirect effects they send rippling through food webs.[3] This enhances bottom-up processes, such as nutrient cycling, and habitat for all sorts of species, including songbirds, butterflies, lizards, and fish. It increases riparian integrity and even improves soil chemistry.

Foreman's childhood experience on a Bermuda beach, where he witnessed a shark attacking and subsequently killing a human, left an indelible impression about the power of large carnivores and their impact on communities. But rather than fearing the wildness they embody, which he refers to by its Middle English name, *wyldeor*, which means "self-willed beast," he embraces it. He explains that wyldeor, like the wolf, which competes with us and frightens us, reminds us that we are not all-powerful. So we killed the carnivores. But in doing so we learned that when we remove them, ecosystems begin to change, and not in healthy ways.[4]

Three elements foundational to rewilding are (1) large protected core reserves, (2) connectivity, and (3) keystone species. These principles form the three *C*'s of rewilding: cores, corridors, and carnivores.[5] Additionally, the rewilding argument rests on three scientific concepts: (1) trophic cascades initiated by top predators maintain ecosystem structure, resilience, and diversity; (2) wide-ranging predators require space; and (3) connectivity between core reserves is essential to meet predators' needs for dispersal and genetic diversity, by creating metapopulations. Thus the keystone species concept provides the central tenet for the rewilding vision.[6] At the time Soulé, Terborgh, Estes, Foreman, and others first formulated this scheme, little evidence existed of terrestrial trophic cascades. Terborgh's work in Lago Guri, Venezuela, provided the most compelling early evidence of ecosystem-scale effects driven by top predators in terrestrial landscapes.

Rewilding called for making a leap of faith and taking ideas developed at local scales to a continental scale. Foreman recalled: "We tried to look at the continent of North America through the feet of a wolf, grizzly bear, or jaguar. Where are the places that could be rewilded and turned into places where these creatures could range for thousands of miles?" They identified four continental wildlife linkages that provide the basis for protecting and restoring landscape permeability in North America: the Boreal Wildway, across the top of the continent; the Pacific Wildway, down the West Coast to Baja California; the Western

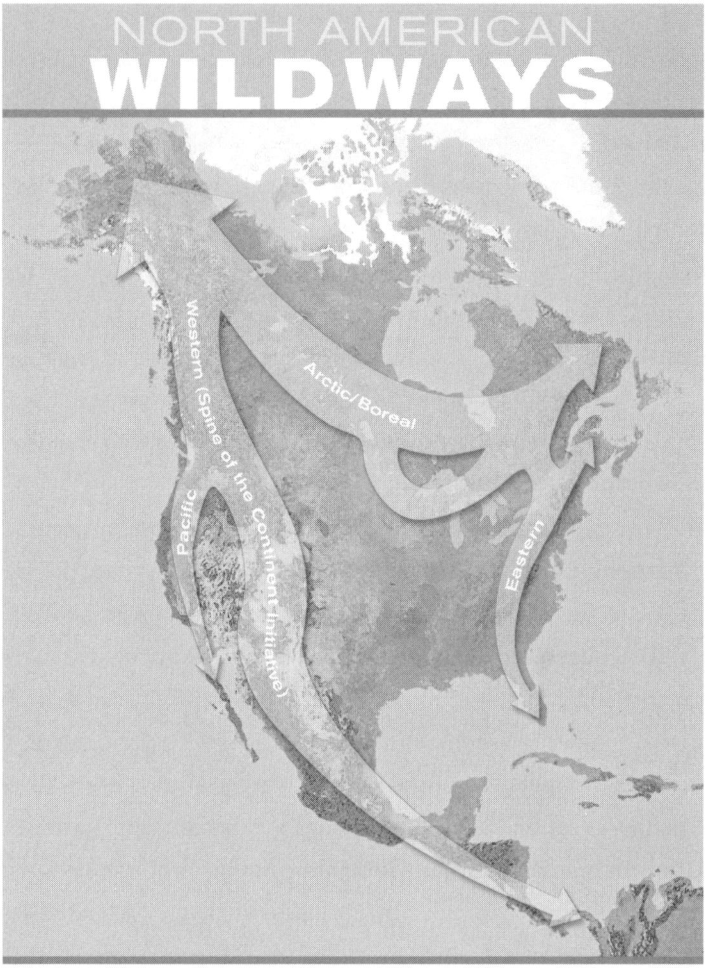

Figure 8.1. North American wildlife linkages

Wildway, also known as the Spine of the Continent Wildway, along the Rocky Mountains and the Sierra Madre Occidental; and the Eastern Wildway.[7]

However, core wild areas and wildlife linkages are not enough. Public and private lands managed for compatible resource and recreational use are essential to provide habitat and dispersal connectivity for many species. Of course, this has limits. Foreman pointed out that not all areas are suitable for rewilding. I recently attended a wildness symposium in Iowa, of all places, which has the lowest proportion of wildlands of any state in the Union (3 percent). As my plane descended toward the Des Moines airport, I looked out the window and saw tidy fenced farms set within a matrix of roads extending from horizon to horizon. I had trouble envisioning free-roaming large carnivores in this environment and could easily imagine the issues that might arise if wolves (*Canis lupus*) returned. Never mind that the white-tailed deer (*Odocoileus virginianus*) population in this state was eating itself out of house and home.

The Wildlands Project stepped forward to address these challenges by encouraging landowners to participate in voluntary actions to protect wildlife linkages and native species. This includes taking advantage of federal and state programs that pay landowners for conservation of their lands, making easements to land trusts, and adjusting land management to protect ecological property values. Since its inception the Wildlands Project has expanded its original concept to encompass large-scale, on-the-ground implementation of its conservation plan, with many partners. To better reflect this effort, in 2009 the organization changed its name to Wildlands Network.

Leopold wrote, "One of the penalties of an ecological education is that one lives alone in a world of wounds. Much of the damage inflicted on land is quite invisible to laymen. An ecologist must either harden his shell and make believe that the consequences of science are none of his business, or he must be the doctor who sees the marks of death in a community that believes itself well and does not want to be told otherwise."[8] Such is the landscape we inhabit in this time of economic uncertainty and climate variability. And out of such uncertainty can arise hope.

In January 2009, Wildlands Network convened the Western Conservation Summit, gathering key people from Canada, the United States, and Mexico to address these issues. The group met in an ancient redwood forest to find common ground and long-term solutions to conserving biodiversity and the web of life on a continental scale. The summit brought together people from diverse perspectives and worldviews who have not always agreed about the best ways to go about saving nature. They came from business, law, education, government service, science, and arts backgrounds and included scientists such as Paul and Anne Ehrlich, Michael Soulé, and E. O. Wilson. Under the sheltering canopy of a grove of giant redwoods they discussed biodiversity, what it means to care for a place, and how to weave together stewardship, science, and natural history to restore landscapes. Grounded by a foundation of the best available science, they avowed the value of keystone predators in maintaining the diversity of life via trophic cascades, the importance of conservation networks that emphasize core habitat areas as well as corridors, and our dependence on the earth's services in the form of ecosystem integrity and resilience.

In crafting their model they focused on Wildland Network's Spine of the Continent Initiative, a campaign to conserve the 5,000-mile-long wildlife corridor that extends from Alaska to Mexico along the Rocky Mountains. This cordillera spans the Arctic to the subtropics, what most people think of as the Wild West, and contains everything from polar bears (*Ursus maritimus*) to Mexican wolves (*Canis lupus baileyi*) to jaguars (*Panthera onca*). For this pathway Soulé and his colleagues identified their objectives as building capacity for conservation and collaborating to connect areas within a compatible-use matrix that includes public and private lands. As if led by the line on the map made by the Rockies and the Sierra Madre, they arrived at consensus, reaching beyond barriers to find common ground. Acknowledging that much had been done by each of the organizations present, they agreed there is much more yet to do. Climate change gives greater urgency to this work, as do issues of social and ecological justice such as the border wall under construction between the United States and Mexico. Those present expressed faith that this conservation dream, though grand, can actually be accomplished.

As their time together drew to a close, participants found hope that their work will help create a world more resilient and whole for our children. What they termed a legacy project, one in which they find a unified voice, will foster diverse, healthy landscapes in this time of global change. Finally, they came away with awareness that out of recognition of the ecological wounds we live with, we will be able to heal the earth.

We have made much progress in large carnivore restoration since Soulé and his colleagues first proposed the rewilding. Back in the 1990s Soulé and Reed Noss wrote, "The greatest impediment to rewilding is an unwillingness to imagine it."[9] Yellowstone National Park provides a shining example of what can happen when one rewilds an ecosystem, with wolf presence related to increased species richness. In February 2008 the wolf was removed from the Endangered Species List in the northern Rockies.[10] This event appeared to be turning the rewilding vision, which some thought outlandish at first, into reality. However, sometimes it feels as if for every two steps forward we take one step back. Wyoming's wolf plan enabled ranchers and others to kill wolves outside protected lands upon delisting. And they proceeded to do just that, reducing the wolf population in that state below a genetically viable number. Consequently, in September 2008 the federal government relisted wolves in the northern Rockies because this population did not meet Endangered Species Act (ESA) criteria for wolf recovery to ensure healthy genetics without linkages between Idaho, Montana, and Wyoming subpopulations. Several more delisting attempts were legally countered by conservation groups. Wolves were successfully delisted in Idaho and Montana in February 2009, and the first wolf hunt took place in these states in fall and winter 2009–2010. The hunt raised questions about the wisdom of allowing animals radio-collared for research purposes to be harvested. Although we have come a long way, we still have far to go.

Integrating Trophic Cascades Science and Ecosystem Management

Returning keystone species to North America represents an effective approach to conservation. And while large carnivore restoration offers formidable challenges, the Yellowstone wolf reintroduction demonstrates that this is feasible. Rewilding addresses our need for an extensive network of public and private wildland reserves. Ecosystem management provides tools to maintain biodiversity and ecosystem services in a broader variety of landscapes, especially those dominated by multiple human uses. The two perspectives differ philosophically, but together they offer a strategy for how trophic cascades science can be used to conserve whole ecosystems and how wildness and civilization can coexist.

Ecologists examine how communities are structured and how they function, questions relevant in conserving both protected ecosystems, such as national parks, and those subject to resource extraction, such as national forests. With the world human population at 6.8 billion and likely to continue to increase by 1.2 percent annually, it is projected to grow to 8–9 billion by 2050.[11]

Human-caused changes to ecosystems will continue apace, as will climate variability. Human activities have caused eutrophication of lakes, acid rain pollution, erosion due to land use and poorly constructed roads, introduction of exotic species, and habitat fragmentation. To help create healthy ecosystems we need to identify what holds them together—the top carnivores—the direction and strengths of their interactions, the components of a system most sensitive to change, and breaking points beyond which restoration may not be possible.[12] Additionally, we need to put human needs into the equation. All the conservation plans and best intentions won't work if we don't figure the human factor into them. Achieving sustainability of biodiversity and whole ecosystems—for humans as well as for wildlife and the habitat they require—involves integrating science, management, and policy.[13] The methods I describe next provide conservation and resource management road maps; they can take us far along what is often a bumpy road with unexpected curves and challenges.

THE CONCEPT of ecosystem management as public policy builds on Aldo Leopold's land ethic philosophy, which grew out of years of working on conservation problems having to do with resource extraction, ungulates, and predators and presented a new concept of land as an organism. His words "A thing is right when it tends to preserve the integrity, stability, and beauty of the biotic community. It is wrong when it tends otherwise"[14] have been cited often as the essence of the land ethic. This passage's spare wording belies the fact that it resulted from a lifetime afield, reflections on his mistakes, and awareness that every living thing plays an ecological role and has ecological value. When he wrote these words he was deeply immersed in educating the public and land managers about the fundamentals of ecology, as applied especially to a Wisconsin deer irruption. He argued passionately that resource management had to shift from a focus on economics and extraction to a more holistic view of what he began referring to as the biotic community—a view that called for maintaining and restoring *all* species, including large predators.[15]

Ecosystem management arose in the early 1980s in response to natural resource issues and was shaped by ecologists and forestry experts such as Jerry

Franklin of the University of Washington and K. Norman Johnson, Thomas Spies, and Frederick Swanson at Oregon State University. However, it did not come into wide use until 1992, when it became the focal point of the US Forest Service's New Perspectives program.[16] Although this approach was created within forestry, it has been applied widely to other resources, such as fish and game, by several agencies. Former Forest Service official and wildlife ecologist Hal Salwasser, inspired by Aldo Leopold's ideas as well as the work of Aldo's son Starker, a wildlife ecologist and policy expert who did much of the groundwork for the Endangered Species Act and large carnivore conservation, led a diverse group of agency scientists and managers in codifying ecosystem management into government policy. According to Salwasser, from the start ecosystem management

> included humans inside the system, which is the way Aldo Leopold thought about it when he started developing his concept of land. By 1920 he clearly understood that the human enterprise had to depend upon using natural resources. And if they did that in ways that diminished the capacity of a place to sustain the yield of those resources, the human enterprise wasn't going to be as well off as otherwise. The main thing that has changed over time has been our ecological understanding of a system as being dynamic and that uncertain systems are not equilibrial. The other thing that has changed is that the human enterprise has gotten about three or four times bigger than when Leopold was starting to do his thinking.[17]

Scientists and managers saw this as an exciting, progressive initiative—one that would incorporate cutting-edge ecological knowledge with the more traditional extraction policy in place since the early 1900s, when Gifford Pinchot was the first chief of the Forest Service. Pinchot saw forest management as driven by the need to provide "the greatest good of the greatest number in the long run."[18] He crafted his management model in an environment where Americans were just becoming aware that their resources would not last forever and that unsustainable harvest would eventually lead to a timber famine that would profoundly affect future generations. The Forest Service carried out Pinchot's conservation strategy, with refinements on sustained yield through the 1960s.

In the early 1990s, because of changing social values, new ecological findings, and awareness of the limits of our resources, the Forest Service began to shift from a focus on commodity production to management of forest ecosystems for sustained yield *and* ecological integrity. This was a challenging time for the agency, with timber harvest dropping for a variety of reasons, including conservation of endangered species endemic to old growth, such as the northern subspecies of spotted owl (*Strix occidentalis caurina*), as we saw in chapter 5. Jerry Franklin's and Jack Ward Thomas' revolutionary scientific findings about the value of downed wood, retention of standing dead trees, and maintenance of owl habitat fueled what came to be known as the "new forestry." Salwasser, Johnson, and others rose to the occasion and created an innovative policy that would better address complex ecological and sociological problems—one still in place and evolving today. From a scientific perspective, ecosystem management as policy was new. However, it was rooted in ideas that had been around since Leopold's time and perhaps earlier about land health, stewardship, and living rightly with nature.[19]

The promise of ecosystem management, back in the 1990s and today, is that it can help us meet human needs for resources *and* help conserve ecosystems. That's a lot to take on in one policy statement, but Salwasser and his colleagues gave it their best effort. Like the term *keystone species*, it quickly became popular, and agencies, corporations, and private conservation organizations adopted it. In the process multiple definitions developed, each reflecting a different constituency and worldview. The values inherent in ecosystem management became the subject of debate, and a series of journal articles published in the mid- to late 1990s explored various definitions of this term. These definitions warrant scrutiny because of the different approaches to conservation they represent.

Salwasser defined ecosystem management as the idea that "knowledge and technology can be used in actions to encourage desired conditions of ecosystems for environmental, economic, and social benefits."[20] Conservation biologist R. Edward Grumbine called it "integrating scientific knowledge of ecological relationships within a complex sociopolitical and values framework toward the general goal of protecting native ecosystem integrity over the long term."[21] The Environmental Protection Agency suggested that it meant "to restore and maintain the health, sustainability, and biological diversity of ecosystems while supporting sustainable economies and communities,"[22] and Jerry Franklin defined it simply as "managing ecosystems so as to assure their sustainability."[23]

Generally, ecosystem management entails a Leopoldian environmental ethic and view of ecosystems as being more than the sum of their parts. It differs from its sustained-yield predecessor in that the latter is derived from a Clementsian view of ecosystems as predictable and primarily regulated by laws of orderly community succession, while the former is based on a more progressive Eltonian view of ecosystems as dynamic, complex, and filled with uncertainty.[24] According to policy expert K. Norman Johnson, coauthor and architect of the Northwest Forest Plan,

> we can come up with different definitions, but you really are not talking about a management system. Basically, the horse is out of the barn running free, and you can say you're still in control, but you're not. And so managers now have a very different role than they used to. There's less control; they are less in charge. Whether they are on public or private lands, or out on the seas, there are a lot of other forces that are going to impact them. And there are a lot of demands and consideration of values that they didn't deal with before. Ecosystem management is based on uncertainty, and that is another reason why it is uncontrollable. Management isn't a puzzle, which is how we used to approach it; it's a mystery. So how do you cope with a mystery?[25]

Since its inception ecosystem management has at times lived up to its promise, but it has suffered from overly broad application and misinterpretation. Although in the 1990s it was sometimes seen as the latest fad in conservation and management of natural resources, it has matured. And its basic principles are even more relevant today. A committee of the Ecological Society of America evaluated the scientific basis for ecosystem management and found it at least as much about managing humans as about managing natural resources.[26] Salwasser agreed with this interpetation:

> The term is, if you take both words literally, almost an oxymoron, because management implies that you are influencing something to control the outcome, and

> ecosystems are not controllable. So to me it's the process of applying ecological thought to the management of places and people. Ecosystem management has to address the reality that people's livelihoods depend upon natural resources. But it tries to be cognizant of the full range of complexity and the significance that diversity in both species and processes has in sustaining the ability of those places to be resilient or productive over a long scale.[27]

At its core ecosystem management is based on values, and, as such, as a concept it remains fluid, changing with the times to better reflect the needs of broad constituencies, which include humans, flora, fauna, and whole landscapes. As Jerry Franklin explains,

> it's a multifaceted concept. And obviously, one of the things that it implies is a very holistic approach to a system rather than one that's focused on a single element or outcome. From an ecological point of view, ecosystem management has to incorporate the notion of sustaining diverse ecosystem processes. Additionally, it needs to address human values or objectives. This makes it very much a value-based concept, which reflects our evolving human perspectives about these systems and the fact that we appreciate that we have to pay attention to the various multiple goods and services we get from them, and also sustain those.[28]

Policy expert Robert Lackey suggests that ecosystem management reflects a stage in the continuing evolution of social values and priorities; it is neither a beginning nor an end. It should maintain ecosystems in the appropriate condition to achieve desired benefits defined by society, not scientists. And although scientific information is important for effective ecosystem management, it is but a single element in a decision-making process fundamentally one of public or private choice.[29]

Frederick Swanson, former principal investigator at the H. J. Andrews Experimental Forest (HJA), believes ecosystem management as it was envisioned nearly two decades ago has not died and will never go away. But it operates in extremely varied contexts, which can change over time. These include industrial forestry, wilderness conservation, and urban ecosystems. The basic idea behind it is understanding how systems work, including flows of stocks of water, nutrients, and pollutants:

> We want to know about species and system vulnerability to tweaks. We can let those phenomena trying to run their course do as much work for us in the long term as possible to get the services we want, rather than fight them and lose, or build vulnerability so that when the system flips, it's a much bigger flip-out. Ecosystem management is about learning how the system is working, including the social part of it, and using that knowledge to achieve our ever-evolving perceptions of what we want. The basic notion applies anywhere. Today we live in a world full of uncertainty—more so than when the concept of ecosystem management was created. It will be very interesting to see how we apply it as we deal with climate change issues, for the whole environmental context may shift. What do we accept; how do we deal with loss? How do we try to channel change to our interests?[30]

Ecosystem function depends on structure, diversity, and integrity. If one of our objectives is to improve ecosystem function, then trophic cascades science can work well to provide information about connections between population dynamics and ecological processes. Food web links between strongly interacting species occur everywhere. All systems can be managed more effectively if we are aware of these links and incorporate this knowledge into new strategies. For example, we have seen how trophic cascades in lakes enable managers to biomanipulate carnivorous fish populations to maintain top-down effects, thereby improving water quality.[31]

Trophic cascades science has broad applicability to ecosystem management because it is based on phenomena that improve biodiversity and increase ecosystem resilience and persistence over time. Maintaining diversity in ecosystems enhances their ability to recover naturally after disturbance and still provide resources for human well-being. Additionally, this concept fits well within the social aspect of ecosystem management because it's an engaging topic that can easily be understood by the lay public. However, this science is not necessarily accepted and applied by managers.

One of the current challenges in applying trophic cascades science to support ecosystem management has to do with the concepts of *minimum viable population* and *ecological effectiveness*. They spring from very different philosophies about endangered species recovery. Viability refers to the likelihood of a species to persist for some period of time. Effectiveness refers to the impacts of a species' presence, abundance, and distribution in achieving desired ecosystem conditions. State wildlife agencies adhere to a single-species perspective based on minimum viable population (MVP) sizes—a concept not fully compatible with the holism of trophic cascades science and very different from an ecologically effective population, which represents an ecosystem perspective. For example, the fifteen breeding pairs of wolves specified in Montana's wolf management plan may be a viable population as specified by the ESA, but it may not necessarily be an ecologically effective number of wolves, defined as one capable of triggering a trophic cascade.[32]

Scientists and policy experts suggest that an MVP approach may be insufficient to meet the intent of federal legislation for top carnivores, which affect an ecosystem and the abundance and distribution of other species within it. Rather, we may need to shift to an ecologically effective population perspective, which acknowledges top carnivores' role in structuring ecosystems. James Estes, Michael Soulé, and others have more recently taken the lead in recommending a science-based management approach that incorporates current knowledge that keystone species can be essential to their environment. This strategy may help increase diversity of native species and ecosystem resilience because of keystone effects. For example, an ecologically effective sea otter population creates a shift from a denuded forest floor to a robust kelp forest. In an aspen community an ecologically effective wolf population creates a phase shift from an intensely browsed forest understory to one characterized by sapling growth above browse height. Ecologically effective densities are relative because ecosystems are complex and vary across time and space. However, scientists can measure phase states such as these using standard vegetation sampling methods. Applying the concept of ecological effectiveness to management will call for identifying the processes used to characterize a trophic cascade; defining a functional relationship between these processes and desired outcomes; and developing a quantitative protocol to determine whether an ecosystem or region supports an ecologically effective keystone population. Within this framework, persistence of a desired phase state, such as a thriving kelp forest, would define a recovered population of a listed keystone species.[33]

Taking an ecologically effective approach might result in establishing a higher threshold population for wolves before delisting occurs. And once a species has been delisted, this may cause the Forest Service, which manages most wolf habitat in the northern Rockies, to go beyond its recent focus on maintaining habitat for species to managing keystone species to ensure ecologically effective population levels.

The first lines of the Endangered Species Act state that the purpose of the act is "to provide a means whereby the ecosystems upon which endangered species and threatened species depend may be conserved." Therefore, setting recovery levels based on ecologically effective populations would seem a possible interpretation of this clause, where a species can have a strong effect on the vitality and diversity of an ecosystem.[34]

The biodiversity clause of the National Forest Management Act of 1976 opens the door for this sort of paradigm shift. The clause requires the Forest Service to "provide for diversity of plant and animal communities based on the suitability and capability of the specific land area in order to meet overall multiple use objectives and within the multiple use objectives of a land management plan adopted pursuant to this section, provide, where appropriate, to the degree practicable, for steps to be taken to preserve the diversity of tree species similar to that existing in the region."[35] Salwasser suggests amending this clause to read, "In planning for diversity of plant and animal species and communities to meet overall multiple use objectives, document the estimated effect of each alternative on how keystone species populations and overall distribution and abundances of indicator habitats will sustain diversity."[36] Such revision would have been seen as radical as recently as five years ago. But with global climate change accelerating, along with human modification of ecosystems and overexploitation of resources, perhaps it's time to consider such a change.[37]

Gray wolves can trigger trophic cascades in forest and riparian ecosystems of the northern Rockies, raising questions about the adequacy of the current MVP approach to setting recovery and viability population levels for the species. Although we have much to learn about how these ecosystems function, sufficient scientific evidence exists about wolf influence on these landscapes to warrant reassessing our approach to wolf conservation and ecosystem management. Such a reevaluation could lead to broader changes in current approaches to conservation of highly interactive species. In practice it could result in major modifications in national forest management and would call for a new working relationship between the Forest Service and wildlife agencies.[38]

Trophic Cascades and Adaptive Management

Given unknowns, uncertainty, and changing knowledge and social values, ecosystem management must be adaptive to be sustainable. Adaptive resource management is a science-based approach that reflects current understanding of wildlife principles as well as the needs of broader constituencies, such as ranching communities. C. S. Holling first proposed this progressive approach in the 1970s, with further refinements by Carl Walters and Kai Lee in the 1980s and 1990s. Unlike the command-and-control methods employed by agencies in the past, it allows action in the face of uncertainty by implementing strategies that can be tested and revised as needed. This enables managers to improve policies and create more resilient systems. Described as "learning by doing," adaptive management presupposes that scientific knowledge is in an ongoing state of development. Its steps include the following:

1. Acknowledging uncertainty
2. Quantifying management goals on the basis of core values stakeholders can agree on
3. Identifying pertinent competing biological theories and management recommendations
4. Developing testable hypotheses about policy success
5. Searching for and using information to test hypotheses
6. On the basis of research and experience, selecting what is initially believed to be the best management strategy and implementing it
7. Monitoring management efforts

8. Updating management actions as needed on the basis of theories and models that worked best and on periodic fair-minded review[39]

As conceptualized by Holling, this can be a cumbersome process. Johnson and his colleagues recommend an approach that is more streamlined but that acknowledges uncertainty, continues to view policy and management decisions as testable hypotheses, and employs cost-effective monitoring programs that produce results quickly.[40]

Adaptive management does not require consensus, but it requires openness to change. Theoretically it is a self-correcting method, provided managers are willing to admit when previous strategies have failed. Our human egos and fundamental need to be right make adaptive management inherently challenging to apply. For it to succeed we need to be open to new organizational structures; this calls for self-examination and flexibility.[41] Additionally, because it also requires working with stakeholders, it does not provide a quick fix to any natural resources problem. When applied to habitat and biodiversity issues pertaining to trophic cascades, it takes into consideration that keystone restoration goes far beyond science and incorporates social and political effects.

Swanson was one of the scientists at the center of the scientific findings about the value of old-growth forests and the resulting vast policy and management changes in US national forests in the 1980s and 1990s. These changes included the rise of adaptive management districts in the Pacific Northwest, where he works. He has experienced this approach's full spectrum, from rigorous hypothesis testing to managers making statements such as "Oh, yeah, we adapt all the time" but employing a very loose approach. He defines adaptive management as an active collaboration between scientists and managers and points out that this nexus holds many cultural challenges. Achieving rapprochement between these two groups is essential for it to be effective. Managers work on shorter time horizons than scientists and seek the best solution to local problems. Researchers take a longer, broader view and value academic freedom and debate. Managers operate with greater public scrutiny and less freedom to pursue interests. However, scientists and managers may have shared objectives (e.g., learning how natural systems work). Their collaboration can culminate in mutually beneficial research and joint communications with their peers and constituents.[42] Ideally this process also incorporates stakeholders, via public comment or participation in planning committees.

Swanson describes three venues in which he has seen adaptive management succeed at the HJA. The first was exemplified by the northern spotted owl and old-growth wars of the 1980s and 1990s and entailed looking at paradigm-breaking things, such as the importance of coarse woody debris on land and in streams and the complex structure of forests. In the 1990s the Northwest Forest Plan provided a second venue for adaptive management, which operated at the grand scale of land allocations for the northern spotted owl, within the context of the interagency scientific committee studying this species, and the resulting management plan. A third case occurred at the local scale of the individual adaptive management areas, which have succeeded in sustaining and monitoring old-growth forest. Examples include the ten-year monitoring process for the spotted owl, some watershed work, and considerable improvements to forest conditions.[43]

Within adaptive management, trophic cascades science presents a sound basis to answer questions such as what type of carnivore population can facilitate optimal ecosystem function. This science is particularly useful because many studies in this field have been crafted using a comparative experimental design. Close manipulation of variables (see, for example, the work at Lake Mendota, Wisconsin) enables managers to gain a better understanding of how ecosystems work. Additionally, new research in aquatic and terrestrial systems is defining what constitutes an ecologically effective population of a keystone species and can thus produce ecosystem benefits, such as enhanced nutrient cycling and diversity across many food web levels. Managers can measure these shifts using standard vegetation sampling methods. Adaptive management offers an ideal framework to identify these thresholds.

Joshua Halofsky, a landscape ecologist with the Washington State Department of Natural Resources, has a background as a trophic cascades researcher. As a state agency scientist, he doesn't set policy but is part of a group creating adaptive management strategies. He believes management and policy can recognize the role that predators (including humans) may have on other trophic levels:

> The largest influence of trophic cascades science on my everyday work relates to how it has shaped my perception of human impacts on the environment. It is now second nature for me to not only think about the direct impacts humans may have on a forest stand or a larger landscape, but also the indirect impacts these changes may have on other trophic levels. For example, I recommend that studies of northern spotted owl responses to stand thinning also examine their prey and the vegetation prey depends upon for survival. Therefore, I am interested in owl studies that look at the behavioral response of prey to changes in forest structure, in addition to numerical responses. Rather than focusing only on threatened species, I also think about overall biodiversity, since we learned from food web theory that by increasing biodiversity, we can create a more resilient trophic network.[44]

As an agency with both economic and ecological responsibilities, the Washington State Department of Natural Resources would never actively seek to reintroduce a predator, because of the additional restrictions it would create for managers. Yet wolves have begun to recolonize Washington's North Cascades, creating an opportunity to use trophic cascades science to inform management in places where elk had irrupted prior to the wolves' return. For example, herbivory can limit biodiversity, since herbivores are more attracted to some species (red cedar) than others (spruce). Thus managing to create an ecologically effective wolf population may help improve cedar recruitment. In no-wolf areas, Halofsky proposes creating impediments that may cause a shift in herbivory away from problem areas. These examples illustrate that management informed by trophic cascades science holds promise for the future.

WHAT SCIENTISTS are learning about trophic cascades can be used to craft plans to effectively restore ecosystems. But Aldo Leopold knew that science can take us only so far. Conservation in practice calls for some scientists' willingness to be outspoken, to consider innovative proposals that might initially seem outrageous, such as wolf reintroduction and rewilding, and to find ways to engage people's hearts and minds. Effective conservation means restoring all the pieces of a simplified ecosystem. It calls for applying our best science about how food webs function within an ecosystem management framework. And above all it requires commitment, passion, and connection to place—what Leopold referred to as the land ethic.

Epilogue: Lessons from 763

I cradle her in my arms, keeping the sleeping wolf warm against the morning frost, the way I held my daughters when they were small. However, anything but small, wolf number 763[1] is a sturdy, eighty-pound lactating female. It's mid-May, denning season in the northern Rockies, with the snow only recently melted; she has not yet begun to shed her luxuriant winter coat. I help state wolf specialist Kent Laudon fasten a radio-collar around her neck, fitting it carefully, smoothing her thick, black fur so it doesn't become tangled in the leather strap. I smile, glad for the good fortune of having a young alpha female step into our trap. These are the best wolves to collar, the ones who can tell you most about the pack: the ones most likely to live long, stick around and not disperse, taking your collar and data along with them.

We trapped her in Glacier National Park, Montana, in an aspen stand next to Johnson Meadow, where wolves have been denning since the mid-1980s. Covering ten square miles and permanently closed to the public to protect the wolves that den here, the meadow holds several aspen stands and a long-abandoned homestead. The aspens' pale green leaf buds were just beginning to open that morning as I worked with Laudon and his field crew, all of us coordinating our efforts to handle 763 as gently and efficiently as possible. She stood there, trapped but watching us impassively as Laudon approached with a needle loaded with the anesthetic that would enable her to sleep through what would follow. He deftly jabbed the wolf's thigh and then we waited for fifteen long minutes for the drugs to work. She lay down gradually, wobbling a bit, her head going down last. When she was completely out, he released her foot from the padded leghold trap. Only as he carried her to the tarp where we would work on her did we notice her swollen nipples. She had pups nearby. We blindfolded her and spoke in whispers so as to not disturb her sleep as we drew blood, checked her sharp, white teeth and her body condition (which was superb, even for a new mother), and listened to the drum of her strong, steady heartbeat. I estimated her age: three years old and in her prime.

The collar always goes on last. After we finish fastening it, I continue to hold her, waiting for the anesthetic to begin to wear off, keeping her warm with my body because the drugs we have given her impair her ability to regulate her body temperature. I lower my head and smell her fur, which carries the sweet, clean scent of woods and river and deep, deep wildness—a wildness so primal I have no words for it. Humans have long feared what we can't understand. Part of my work as a scientist involves trying to understand wildness. I pick up one of her front paws and hold it flat against my palm. It is bigger than my outspread hand.

She takes a long time to awaken. Eventually I feel the smooth muscles over her ribs contract. One of her ears twitches, reacting to some sound audible only to her. We ease her off the tarp and onto a soft bed of new spring grass. As we prepare to waken her fully I offer thoughts that she live long and well, that she birth many strong pups in her lifetime, filling this meadow for many generations with their howls and wildness, and that she steer clear of trouble. She has an official number, but before we release her I give her another name. I call her Nina, after my mentor Nina Leopold Bradley, who has helped deepen my understanding of her father's work and the ecological value of wolves. Laudon injects the wolf with an antidote that will reverse the anesthesia completely. Within five minutes the wolf rises, turning back to look at us as she leaves. Her amber eyes burn into mine before she vanishes into the shadows at the edge of the meadow.

I thank the wolfen Nina for the data she will give me, which will illuminate keystone species effects. Her collar's steady stream of high-quality location data—which produce one fix every three hours, with an error of plus or minus one meter (about one yard)—will tell me the story of how she is influencing elk behavior and how this in turn is shaping patterns in this landscape. She will show me how wolf presence affects everything, from the way aspens grow to songbird diversity to mycorrhizal fungi.

Four months later it is August, and I am in an old-growth larch forest surveying Nina's den, which she dug last spring into a south-facing hillside, about a mile and a half from the ancestral den this pack has used for all but a handful of the past twenty-three years. Her collar data suggest the pack is no longer near either den. It is one of those glorious late summer days in the Rockies, with sunlight shafting through the forest, silhouetting the blown seed heads of fireweed and grasses, the larch needles just beginning to turn. I shed layers, enjoying the gift of this day. As I measure vegetation I catch a quick black blur in a nearby huckleberry thicket—a wolf, checking us out. I try radio-tracking but fail to pick up a signal. My field crew and I hurry to finish. This encounter bothers me because I make every effort to have no contact with wolves as I gather plant data, to avoid disturbing them any more than necessary. It may seem paradoxical that someone who collars wolves would avoid encountering them, but it is one way I try to minimize the impacts of my research on this community.

We pack out our field equipment and bushwhack away from the den, moving west into the meadow, traversing it on a labyrinth of wolf trails through waist-high, ochre grass. Eventually we find the primary wolf trail, which is wider than the rest and meanders through this landscape like a river. Johnson Meadow seems no less primordial than I recall, with bones everywhere and wolf lays—patches of flattened grass marked with wolf scats—indicating where the pack has rested. My companions on this fine day include Glacier National Park carnivore biologist John Waller; fire ecologist Dan Donato; Dan's wife, Melanie Stidham, a forest ecologist; and my field technicians Craig DeMars and Neal Wight. Waller is here to learn more about my research and to offer advice; Donato, to help me parse out the meadow's fire ecology and differentiate it from wolf effects on the aspens.

All at once, just inside the conifers hemming the meadow to the south, we see a wolf, and then another, and another—seven in all, mostly black. Wolf pups, now almost the same size as adults, but nevertheless identifiable as pups because of their awkward movements and curiosity. They lack the elegance and spare movements of their elders. They coyly peek out at us from behind lodgepole pines too absurdly spindly to provide much cover. We stop in our tracks and watch in astonishment. Waller and I exchange looks. One of the pups beds down in a patch of violet asters, fifty yards from us. One of its siblings follows suit. Another one starts chasing a butterfly. Then it dawns on us that they are not at all disturbed by our presence. Waller nods. We watch them silently for a few minutes and then continue on our way, weaving through the tall grass, leaving a wake of leaping grasshoppers. As I walk on, I reflect on the day I collared their mother, how I held her to keep her warm, how she must have returned to her den that day smelling of me. And I realize the pups and I are no strangers to one another.

We move toward the ancestral den, located in a stand of aspens that graces a low knoll, to examine how a trophic cascade looks in an area of such high wolf use. When we get there we sit at the edge of the aspens amid the detritus of wolf life—deer bones cracked in two by the wolves' sharp carnassial teeth, the marrow thoroughly sucked out, a mauled plastic soft drink bottle the pups used as chew toy. The aspens' coinlike leaves, tinged yellow, shimmy in the breeze as we discuss top-down (wolf) versus bottom-up (fire) effects on the trees in this meadow. I point out the classic recruitment gap here, with lots of old aspens, no middle-aged aspens, and thousands of young trees of various ages coming up straight trunked, many reaching above browse height. Donato comments that the saplings' multiple age classes show a distinct release from herbivory and continuing recruitment, rather than the even-aged flush of growth characteristic of fire response. I note abundant signs of ungulate presence—piles of elk and deer pellets deposited in the grass and among the aspens—even though this place represents the inner sanctum of wolf activity in the North Fork.

We eat lunch and talk about how all things are connected, and that events that occurred nearly a century ago, when this site was homesteaded, are plainly written on the landscape today: a midden heap, wolves removed by settlers, the resulting explosion in elk numbers, leaving ecological artifacts such as the missing aspen age classes and the heavily scarred bark on the older trees. We talk about the stories housed in this meadow—the history of enough wolf lives to repopulate northwestern Montana, stories that substantiate what Aldo Leopold taught us so long ago about the significance of saving all the pieces. This meadow shows how many of the habitat degradation and extinction processes precipitated by the elimination of wolves ninety years ago are reversible. Experiencing landscapes like this, where keystone predators rule and all ecological functions driven by them are present freely, inspires one to try to sustain these interactions beyond protected federal lands.

Trophic cascades and ecosystem management are ultimately about relationships—between humans and predators and landscapes. Maintaining and tending these relationships in our own communities requires commitment and vision. In this book I touched on the work of some of the scientists who are providing us with evidence to sustain this vision. As one of the many people currently studying trophic cascades, I am constantly astonished and humbled by the complexity and beauty of food web interactions I observe, such as the ones in Johnson Meadow. Becoming aware of these relationships is integral to advancing trophic cascades science. Putting this awareness to work to help create a more sustainable relationship between humans and the earth represents a leading challenge for science and conservation.

According to Michael Soulé, "trophic cascades science provides copious evidence for concluding that the unpredictable, devastating, downstream effects of apex predator removal, particularly herbivore and mesopredator release, are major drivers of global ecological collapse."[2] Strong words. Yet these cascades continue to be controversial in scientific and management circles. While most scientists accept their existence, many are caught up in scholarly arguments about the details of how they operate, the validity of the keystone concept, whether top-down or bottom-up forces prevail, and whether one can make cross-ecosystem generalizations about these phenomena. These arguments are highly relevant; however, because of all we risk losing, it is important for policy makers to move beyond them and begin to apply trophic cascades principles to conservation and natural resources management.

In the century since we created the science of ecology, we have learned and generally agree that ecosystems are complex and multicausal. We also agree that the world is changing rapidly as a result of climate variability and the human population explosion. John Terborgh and James Estes consider the loss of top predators from ecosystems worldwide due to persecution by humans a crisis as significant as climate change. They suggest that as with climate change, the problem of loss of biodiversity caused by loss of keystones will only worsen until we take measures to correct it.[3] While restoring keystones will not solve all our problems, it will create more resilient systems and help restore and maintain biodiversity. The research and ecological restoration efforts I have presented in this book provide a wellspring of ideas for how to accomplish this.

I CONTINUED my den survey that August, working into September at the older, historical den in the center of Johnson Meadow. I never saw the pups again. Now it is late autumn, the aspens have shed their leaves, and a scrim of fresh snow dusts the mountains. The memory of that late summer day, the pups watching us and us watching them back in wonder, comes to me as I analyze my data. The columns of numbers on my computer screen ineluctably show the mark of the wolf's tooth on the meadow's ecology: how the wolves' return in the mid-1980s has changed so much. These data demonstrate how wolves are an essential part of the web of life. Wolf presence has indirectly made aspens and shrubs grow again after many years of overbrowsing by ungulates and has improved habitat for songbirds and elk. And five months after I collared Nina, the data from her collar are coming in clear and true, providing a line of incontrovertible empirical evidence to support my vegetation and songbird data. Predictably, her collar data spin the story of what it means to be a wolf and how her wildness can mend the web of life and enable other species to live well. She is a keystone: the shaping force that reforms ecosystems into wholes. And her revelations fill me with hope.

Notes

Introduction. Visitors from the North

1. Colorado Division of Wildlife, "Guidelines for Response to Gray Wolf Reports in Colorado" (Denver: Colorado Division of Wildlife, 2005).
2. Curt Meine, *Correction Lines: Essays on Land, Leopold, and Conservation* (Washington, DC: Island Press, 2004), 117–32.

Chapter 1. Patterns in an Ecosystem

1. Robert Ream et al., "First Wolf Den in Western United States in Recent History," *Northwestern Naturalist* 70 (1989): 39–40; the name of the meadow is fictitious to protect the privacy of the wolves that den here.
2. My study, Trophic Cascades Involving Humans, Wolves, Elk, and Aspen in the Crown of the Continent Ecosystem, is taking place in Glacier National Park, Montana, and Waterton Lakes National Park, Alberta, Canada. The wildlife aspect of this research is part of the Southwest Alberta Montane Research Program, an interagency transboundary effort in which I am a collaborator.
3. William H. Romme et al., "Aspen, Elk, and Fire in Northern Yellowstone Park," *Ecology* 76, no. 7 (1995): 2097–2106; Norbert V. DeByle, "Wildlife," in *Aspen: Ecology and Management in the Western United States*, ed. Norbert V. DeByle and Robert P. Winokur, General Technical Report RM-119 (Fort Collins, CO: US Department of Agriculture, Forest Service, 1985), 125–55.

4. Robert T. Paine, "Food Webs: Linkage, Interaction Strength, and Community Infrastructure," *Journal of Animal Ecology* 49, no. 3 (1980): 666–85.
5. Robert J. Taylor, *Predation* (New York: Chapman and Hall, 1984).
6. Immediate effects of fire may be detrimental to preferred grizzly food items, such as huckleberries. However, longer-term effects may be positive, including rejuvenation of decadent huckleberry stands in some areas. See James K. Agee, *Fire Ecology of Pacific Northwest Forests* (Washington, DC: Island Press, 1996), 180.
7. Michael Begon, John L. Harper, and Colin R. Townsend, *Ecology: Individuals, Populations, and Communities*, 3rd ed. (Hoboken, NJ: Blackwell Science, 1996), 625.
8. Sharon E. Kingsland, "Conveying the Intellectual Challenge of Ecology: An Historical Perspective," *Frontiers in Ecology and the Environment* 2, no. 7 (2004): 367–74.
9. Charles Elton, *Animal Ecology*, 2nd ed. (Chicago: University of Chicago Press, 2001), 50–70; Donald Worster, *Nature's Economy: A History of Ecological Ideas*, 2nd ed. (Cambridge: Cambridge University Press, 1994), 299.
10. Susan Flader, *Thinking Like a Mountain: Aldo Leopold and the Evolution of an Ecological Attitude toward Deer, Wolves, and Forests* (Madison: University of Wisconsin Press, 1974), 242–43.
11. Aldo Leopold, "Deer Irruptions," *Wisconsin Academy of Sciences, Arts, and Letters* (1943): 351–66; Aldo Leopold, Hunting Journals, ca. 1920–45, Aldo Leopold Papers, University of Wisconsin, Archival Collections, 10-7, 2.
12. Curt Meine, *Aldo Leopold: His Life and Work* (Madison: University of Wisconsin Press, 1988), 368–69; Curt Meine and Richard L. Knight, eds., *The Essential Aldo Leopold: Quotations and Commentaries* (Madison: University of Wisconsin Press, 1999), xvi.
13. Nelson G. Hairston, Frederick E. Smith, and Lawrence B. Slobodkin, "Community Structure, Population Control, and Competition," *American Naturalist* 94, no. 879 (1960): 421–25.
14. James A. Estes, "Carnivory and Trophic Connectivity in Kelp Forests," in *Large Carnivores and the Conservation of Biodiversity*, ed. Justina Ray et al. (Washington, DC: Island Press, 2005), 61–80.
15. William J. Ripple and Robert L. Beschta, "Wolf Reintroduction, Predation Risk, and Cottonwood Recovery in Yellowstone National Park," *Forest Ecology and Management* 184 (2003): 299–313; Mark Hebblewhite et al., "Human Activity Mediates a Trophic Cascade Caused by Wolves," *Ecology* 86, no. 8 (2005): 2135–44.
16. Rolf O. Peterson et al., "Temporal and Spatial Aspects of Predator-Prey Dynamics," *Alces* 39 (2003): 215–32; Brian E. McLaren and Rolf O. Peterson, "Wolves, Moose, and Tree Rings on Isle Royale," *Science* 266, no. 5190 (1994): 1555–58.
17. Mary E. Power, "Top-Down and Bottom-Up Forces in Food Webs: Do Plants Have Primacy?" *Ecology* 73, no. 3 (1992): 733–46.
18. Oswald J. Schmitz, "Predators Have Large Effects on Ecosystem Properties by Changing Plant Diversity, Not Plant Biomass," *Ecology* 87, no. 6 (2006): 1432–37.

19. James A. Estes, Norman S. Smith, and John F. Palmisano, "Sea Otter Predation and Community Organization in the Western Aleutian Islands, Alaska," *Ecology* 59, no. 4 (1978): 822–33.
20. Donald R. Strong, "Are Trophic Cascades All Wet? Differentiation and Donor Control in Speciose Ecosystems," *Ecology* 73, no. 3 (1992): 747–54.
21. Edward O. Wilson, *The Diversity of Life* (Cambridge, MA: Belknap Press of Harvard University Press, 1992), 243–80; J. Emmett Duffy, "Biodiversity and Ecosystem Function: The Consumer Connection," *Oikos* 99, no. 2 (2002): 201–19.
22. Clifford A. White, Michael C. Feller, and Suzanne Bayley, "Predation Risk and the Functional Response of Elk-Aspen Herbivory," *Forest Ecology and Management* 181 (2003): 77–97.
23. Joel S. Brown, William Laundré, and Mahesh Gurung, "The Ecology of Fear: Optimal Foraging, Game Theory, and Trophic Interactions," *Journal of Mammalogy* 80, no. 2 (1999): 385–99.
24. John Terborgh et al., "Vegetation Dynamics of Predator-Free Land-Bridge Islands," *Journal of Ecology* 94, no. 2 (2006): 253–63.
25. Gary A. Polis and Donald R. Strong, "Food Web Complexity and Community Dynamics," *American Naturalist* 147, no. 5 (1996): 813–46.
26. Kyran E. Kunkel et al., "Winter Prey Selection by Wolves and Cougars in and near Glacier National Park, Montana," *Journal of Wildlife Management* 63 (1999): 901–10.
27. Ibid.
28. Walter F. Mueggler, "Age Distribution and Reproduction of Intermountain Aspen Stands," *Western Journal of Applied Forestry* 4, no. 2 (1989): 41–45.
29. Elliot Fox and Kevin Van Tighem, "The Belly River Wolf Study: Six Month Interim Report" (Waterton Lakes National Park, Alberta: Parks Canada, 1994).
30. Scott C. Mills, Michael E. Soulé, and Daniel F. Doak, "The Keystone-Species Concept in Ecology and Conservation," *BioScience* 43, no. 4 (1993): 219–25; Michael E. Soulé et al., "Strongly Interacting Species: Conservation Policy, Management, and Ethics," *BioScience* 55, no. 2 (2005): 168–76.
31. John Terborgh and James A. Estes, preface to *Trophic Cascades*, ed. John Terborgh and James A. Estes (Washington, DC: Island Press, forthcoming).

Chapter 2. Living in a Landscape of Fear: Trophic Cascades Mechanisms

1. Robert T. Paine, interview by Cristina Eisenberg, January 31, 2008, University of Washington, Seattle.
2. Nelson G. Hairston, Frederick E. Smith, and Lawrence B. Slobodkin, "Community Structure, Population Control, and Competition," *American Naturalist* 94, no. 879 (1960): 421–25.
3. William Stolzenburg, *Where the Wild Things Were: Life, Death, and Ecological Wreckage in a Land of Vanishing Predators* (New York: Bloomsbury, 2008), 6–25.

4. Robert T. Paine, interview by Cristina Eisenberg, January 31, 2008, University of Washington, Seattle.
5. Stephen D. Fretwell, "Food Chain Dynamics: The Central Theory of Ecology?" *Oikos* 50, no. 3 (1987): 291–301; Robert S. Steneck, "An Ecological Context for the Role of Large Carnivores in Conserving Biodiversity," in *Large Carnivores and the Conservation of Biodiversity*, ed. Justina Ray et al. (Washington, DC: Island Press, 2005), 10–11.
6. Hairston, Smith, and Slobodkin, "Community Structure."
7. Aldo Leopold, "Deer Irruptions," *Wisconsin Academy of Sciences, Arts, and Letters* (1943): 351–66; Charles Elton, *Animal Ecology*, 2nd ed. (Chicago: University of Chicago Press, 2001), 115.
8. James A. Estes et al., "Killer Whale Predation on Sea Otters Linking Oceanic and Nearshore Ecosystems," *Science* 282, no. 5388 (1998): 473–76.
9. William W. Murdoch, "'Community Structure, Population Control, and Competition,' A Critique," *American Naturalist* 100, no. 912 (1966): 219–26; Stevan Arnold, interview by Cristina Eisenberg, February 7, 2008, Oregon State University, Corvallis.
10. Anthony R. E. Sinclair et al., "Testing Hypotheses of Trophic Level Interactions: A Boreal Forest Ecosystem," *Oikos* 89, no. 2 (2000): 313–28.
11. Gary A. Polis, "Food Webs, Trophic Cascades, and Community Structure," *Australian Journal of Ecology* 19 (1994): 121–36; Gary A. Polis and Donald R. Strong, "Food Web Complexity and Community Dynamics," *American Naturalist* 147, no. 5 (1996): 813–46.
12. Michael E. Soulé (chairman, Wildlands Network), interview by Cristina Eisenberg, October 30, 2008, Paonia, CO.
13. Graeme Caughley, "Eruption of Ungulate Populations, with Emphasis on Himalayan Thar in New Zealand," *Ecology* 51, no. 1 (1970): 56.
14. Lauri Oksanen et al., "Exploitation Ecosystems in Gradients of Primary Productivity," *American Naturalist* 118, no. 2 (1981): 240–61; Fretwell, "Food Chain Dynamics."
15. Oksanen et al., "Exploitation of Ecosystems."
16. Robert T. Paine, interview by Cristina Eisenberg, January 31, 2008, University of Washington, Seattle.
17. Robert T. Paine, "A Note on Trophic Complexity and Community Stability," *American Naturalist* 103, no. 929 (1969): 91–93; Bruce A. Menge et al., "The Keystone Species Concept: Variation in Interaction Strength in a Rocky Intertidal Habitat," *Ecological Monographs* 64, no. 3 (1994): 250; Michael E. Soulé et al., "Strongly Interacting Species: Conservation Policy, Management, and Ethics," *BioScience* 55, no. 2 (2005): 168–76.
18. Joel S. Brown, William Laundré, and Mahesh Gurung, "The Ecology of Fear: Opti-

mal Foraging, Game Theory, and Trophic Interactions," *Journal of Mammalogy* 80, no. 2 (1999): 385–99.

19. Joel Berger, *The Better to Eat You With: Fear in the Animal World* (Chicago: University of Chicago Press, 2008).
20. Aldo Leopold, Shack Journals 1935–1948, Aldo Leopold Papers, University of Wisconsin, Archival Collections, Diaries and Journals, June 17, 1939, 101; June 20–21, 1940, 178; Otis S. Bersing, *A Century of Wisconsin Deer*, 2nd ed. (Madison: Wisconsin Conservation Department, Game Management Division, 1966), 13; Aldo Leopold, *Report on a Game Survey of the North Central States* (Washington, DC: Sporting Arms and Ammunition Manufacturers Institute, 1931).
21. Nina Leopold Bradley (director, Aldo Leopold Foundation), interview by Cristina Eisenberg, September 29, 2004, Baraboo, WI.
22. Joel Berger, "Carnivore Repatriation and Holarctic Prey: Narrowing the Deficit in Ecological Effectiveness," *Conservation Biology* 21, no. 4 (2007): 1105–16.
23. Joel Berger et al., "A Mammalian Predator-Prey Imbalance: Grizzly Bear and Wolf Extinction Affect the Diversity of Avian Neotropical Migrants," *Ecological Applications* 11, no. 4 (2002): 947–60.
24. Richard Kennedy, DVM, e-mail to Cristina Eisenberg, December 24, 2008.
25. Oswald J. Schmitz, Andrew P. Beckerman, and Kathleen M. O'Brien, "Behaviorally Mediated Trophic Cascades: Effects of Predation Risk on Food Web Interactions," *Ecology* 78, no. 5 (1997): 1388–99.
26. Valerius Geist, "Adaptive Behavioral Strategies," in *North American Elk: Ecology and Management* (Washington, DC: Smithsonian Institution Press, 2002), 289–333.
27. Mark Hebblewhite, Evelyn H. Merrill, and Trent L. McDonald, "Spatial Decomposition of Predation Risk Using Resource Selection Functions: An Example in a Wolf-Elk Predator-Prey System," *Oikos* 111, no. 1 (2005): 101–11.
28. Mark Hebblewhite and Dan Pletscher, "Effects of Elk Group Size on Predation by Wolves," *Canadian Journal of Zoology* 80 (2002): 800–809.
29. Joshua Halofsky and William J. Ripple, "Fine-Scale Predation Risk on Elk after Wolf Reintroduction in Yellowstone National Park, USA," *Oecologia* 155 (2008): 868–77.
30. Douglas Smith (director, Yellowstone Gray Wolf Restoration Project), interview by Cristina Eisenberg, November 5, 2008, Yellowstone National Park.
31. Kyran E. Kunkel et al., "Winter Prey Selection by Wolves and Cougars in and near Glacier National Park, Montana," *Journal of Wildlife Management* 63 (1999): 901–10.
32. Stewart Liley and John Creel, "What Best Explains Vigilance in Elk: Characteristics of Prey, Predators, or the Environment?" *Behavioral Ecology* 19, no. 2 (2008): 245–54; Matthew Kauffman et al., "Landscape Heterogeneity Shapes Predation in a Newly Restored Predator-Prey System," *Ecology Letters* 10, no. 8 (2007): 690–700.

33. Durward K. Allen, *Wolves of Minong: Their Vital Role in a Wild Community* (New York: Houghton Mifflin, 1979).
34. Adolph Murie, *The Ecology of the Coyote in the Yellowstone*, Fauna Series no. 4 (Washington, DC: US Government Printing Office, 1940); Douglas Smith (director, Yellowstone Gray Wolf Restoration Project), interview by Cristina Eisenberg, November 5, 2008, Yellowstone National Park.
35. David S. Wilcove, "Nest Predation in Forest Tracts and the Decline of Migratory Songbirds," *Ecology* 66, no. 4 (1985): 1211–14.
36. Michael E. Soulé et al., "Reconstructed Dynamics of Rapid Extinctions of Chaparral-Requiring Birds in Urban Habitat Islands," *Conservation Biology* 2, no. 1 (1988): 75–92.
37. Ibid.
38. Michael E. Soulé (chairman, Wildlands Network), interview by Cristina Eisenberg, October 30, 2008, Paonia, CO; Kevin R. Crooks and Michael E. Soulé, "Mesopredator Release and Avifaunal Extinctions in a Fragmented System," *Nature* 400 (1999): 563–66.
39. H. Valladas et al., "Paleolithic Paintings: Evolution of Prehistoric Cave Art," *Nature* 413, no. 6855 (2001): 479.
40. Anthony D. Barnosky et al., "Assessing the Causes of Late Pleistocene Extinctions on the Continents," *Science* 306, no. 5693 (2004): 70–75.
41. Estella B. Leopold, interview by Cristina Eisenberg, January 30, 2008, University of Washington, Seattle.
42. Paul S. Martin, *Twilight of the Mammoths: Ice Age Extinctions and the Rewilding of America* (Berkeley: University of California Press, 2005), 31–33.
43. Berger, "Carnivore Repatriation and Holarctic Prey."
44. Barnosky et al., "Assessing the Causes of Late Pleistocene Extinctions."
45. Andrea S. Laliberte and William J. Ripple, "Range Contractions of North American Carnivores and Ungulates," *BioScience* 54, no. 2 (2004): 123–38.

Chapter 3. Origins: Aquatic Cascades

1. Edward F. Ricketts, Jack Calvin, and Joel W. Hedgpeth, *Between Pacific Tides*, 5th ed. (Stanford, CA: Stanford University Press, 1992), 216–17.
2. Anne Wertheim Rosenfeld and Robert T. Paine, *The Intertidal Wilderness* (Berkeley: University of California Press, 2002).
3. Ibid., 36.
4. G. Evelyn Hutchinson, "Homage to Santa Rosalia, or Why Are There So Many Kinds of Animals?" *American Naturalist* 93, no. 870 (1959): 145–59.
5. Robert T. Paine, interview by Cristina Eisenberg, January 31, 2008, University of Washington, Seattle.

6. Robert T. Paine and Simon A. Levin, "Intertidal Landscapes: Disturbance and the Dynamics of Pattern," *Ecological Monographs* 51, no. 2 (1981): 145–78.
7. Robert T. Paine, interview by Cristina Eisenberg, January 31, 2008, University of Washington, Seattle.
8. Robert T. Paine, "Food Web Complexity and Species Diversity," *American Naturalist* 100, no. 910 (1966): 65–75; Robert T. Paine, "A Note on Trophic Complexity and Community Stability," *American Naturalist* 103, no. 929 (1969): 91–93.
9. Robert T. Paine et al., "Perturbation and Recovery Patterns of Starfish-Dominated Intertidal Assemblages in Chile, New Zealand, and Washington State," *American Naturalist* 125, no. 5 (1985): 679–91; Robert T. Paine, "Food Webs; Linkage, Interaction Strength, and Community Infrastructure," *Journal of Animal Ecology* 49, no. 3 (1980): 666–85.
10. Bruce Menge, "Indirect Effects in Marine Rocky Intertidal Interaction Webs: Patterns and Importance," *Ecological Monographs* 65, no. 1 (1995): 21–74.
11. Bruce Menge et al., "The Keystone Species Concept: Variation in Interaction Strength in a Rocky Intertidal Habitat," *Ecological Monographs* 64, no. 3 (1994): 249–86.
12. Karl W. Kenyon, "The Sea Otter in the Eastern Pacific Ocean," *North American Fauna* 68 (1969): 1–352.
13. Ibid.
14. Ricketts, Calvin, and Hedgpeth, *Between Pacific Tides*, 98–100.
15. Judith Connor and Charles Baxter, *Kelp Forests* (Monterey, CA: Monterey Bay Aquarium, 1989).
16. James Estes, interview by Cristina Eisenberg, March 31, 2008, Santa Cruz, CA.
17. Ibid.
18. James A. Estes and John F. Palmisano, "Sea Otters: Their Role in Structuring Nearshore Communities," *Science* 185, no. 4156 (1974): 1058–60; James A. Estes, Norman S. Smith, and John F. Palmisano, "Sea Otter Predation and Community Organization in the Western Aleutian Islands, Alaska," *Ecology* 59, no. 4 (1978): 822–33.
19. Estes, Smith, and Palmisano, "Sea Otter Predation."
20. David B. Irons, Robert G. Anthony, and James A. Estes, "Foraging Strategies of Glaucous-Winged Gulls in Rocky Intertidal Communities," *Ecology* 67, no. 6 (1986): 1460–74.
21. James A. Estes, Charles H. Peterson, and Robert S. Steneck, "Direct and Indirect Effects of Apex Predators in Higher Latitude Coastal Oceans," in *Trophic Cascades*, ed. John Terborgh and James A. Estes (Washington, DC: Island Press, forthcoming).
22. David O. Duggins, "Kelp Beds and Sea Otters: An Experimental Approach," *Ecology* 61, no. 3 (1980): 447–53.
23. James A. Estes, personal communication, April 26, 2009.

24. James A. Estes et al., "Killer Whale Predation on Sea Otters Linking Oceanic and Nearshore Ecosystems," *Science* 282, no. 5388 (1998): 473–76.
25. Allan M. Springer et al., "Sequential Megafaunal Collapse in the North Pacific Ocean: An Ongoing Legacy of Industrial Whaling?" *Proceedings of the National Academy of Sciences of the United States of America* 100, no. 21 (2003): 12223–28.
26. American Cetacean Society, *Field Guide to the Orca*, Sasquatch Field Guide Series (Seattle, WA: Sasquatch Books, 1990).
27. Estes et al., "Killer Whale Predation"; Springer et al., "Sequential Megafaunal Collapse."
28. James A. Estes et al., "Continuing Sea Otter Population Declines in the Aleutian Archipelago," *Marine Mammal Science* 21, no. 1 (2005): 169–72.
29. Terrie M. Williams et al., "Killer Appetites: Assessing the Role of Predators in Ecological Communities," *Ecology* 85, no. 12 (2004): 3373–84.
30. Sally A. Mizroch and Dale W. Rice, "Have North Pacific Killer Whales Switched Prey Species in Response to Depletion of the Great Whale Populations?" *Marine Ecology Progress Series* 310 (2006): 235–46; Paul R. Wade et al., "Killer Whales and Marine Mammal Trends in the North Pacific: A Re-Examination of Evidence for Sequential Megafauna Collapse and the Prey-Switching Hypothesis," *Marine Mammal Science* 23, no. 4 (2006): 766–802; Douglas P. DeMaster et al., "The Sequential Megafaunal Collapse Hypothesis: Testing with Existing Data," *Progress in Oceanography* 68 (2006): 329–42.
31. Paul R. Wade, Jay M. Ver Hoef, and Douglas P. DeMaster, "Mammal-Eating Killer Whales and Their Prey: Trend Data for Pinnipeds and Sea Otters in the North Pacific Ocean Do Not Support the Sequential Megafaunal Collapse Hypothesis," *Marine Mammal Science* 25, no. 3 (2009): 737–47; James A. Estes et al., "Trend Data for Pinnipeds and Sea Otters in the North Pacific Ocean Do Support the Sequential Megafaunal Collapse Hypothesis, but Is That the Point?" *Marine Mammal Science* (forthcoming).
32. Robert T. Paine et al., *The Decline of the Steller Sea Lion in Alaskan Waters: Untangling Food Webs and Fishing Nets* (Washington, DC: National Academies Press, 2003), 7–8; National Oceanic and Atmospheric Administration, National Marine Fisheries Service, "Recovery Plan for the Steller Sea Lion: Eastern and Western Distinct Population Segments (*Eumetopias jubatus*)," Revision (Silver Spring, MD: National Oceanic and Atmospheric Administration, National Marine Fisheries Service, 2008).
33. Shauna E. Reisewitz, James A. Estes, and Charles A. Simenstad, "Indirect Food Web Interactions: Sea Otters and Kelp Forest Fishes in the Aleutian Archipelago," *Oecologia* 146, no. 4 (2006): 623–31.
34. Robert G. Anthony et al., "Bald Eagles and Sea Otters in the Aleutian Archipelago: Indirect Effects of Trophic Cascades," *Ecology* 89, no. 10 (2008): 2725–35.
35. James Estes, interview by Cristina Eisenberg, January 27, 2009, Santa Cruz, CA.

36. Daniel C. Donato et al., "Response to Comments on 'Post-Wildfire Logging Hinders Regeneration and Increases Fire Risk,'" *Science* 313, no. 5787 (2006): 615–17; Eric D. Forsman et al., eds., *Demography of the Northern Spotted Owl: Proceedings of a Workshop, Fort Collins, Colorado, December 1993*, Studies in Avian Biology Series, no. 17 (Los Angeles: Cooper Ornithological Society, 1996).
37. Estes, Peterson, and Steneck, "Direct and Indirect Effects of Apex Predators."
38. Raymond McFarland, *A History of the New England Fisheries: With Maps* (New York: University of Pennsylvania, 1911), 12–13.
39. Robert S. Steneck, John Vavrinec, and Amanda V. Leland, "Accelerating Trophic-Level Dysfunction in Kelp Forest Ecosystems of the Western North Atlantic," *Ecosystems* 7, no. 4 (2004): 323–32.
40. Ibid.; Mark D. Bertness et al., "Do Alternate Stable Community States Exist in the Gulf of Maine Rocky Intertidal Zone?" *Ecology* 83 (2002): 3434–48.
41. Ransom A. Myers and Boris Worm, "Rapid Worldwide Depletion of Predatory Fish Communities," *Nature* 423 (2003): 280–83.
42. Ransom A. Myers et al., "Cascading Effects of the Loss of Apex Predatory Sharks from a Coastal Ocean," *Science* 315, no. 5820 (2007): 1846–50.
43. Charles H. Peterson et al., "Site-Specific and Density-Dependent Extinction of Prey by Schooling Rays: Generation of a Population Sink in Top-Quality Habitat for Bay Scallops," *Oecologia* 129, no. 3 (2001): 349–56.
44. Estes, Peterson, and Steneck, "Direct and Indirect Effects of Apex Predators."
45. P. S. Lake et al., "Global Change and the Biodiversity of Freshwater Ecosystems: Impacts on Linkages between Above-Sediment and Sediment Biota," *BioScience* 50, no. 12 (2000): 1099–1106.
46. Stephen R. Carpenter, James F. Kitchell, and James R. Hodgson, "Cascading Trophic Interactions and Lake Productivity," *BioScience* 35, no. 10 (1985): 634–39.
47. John Langdon Brooks and Stanley I. Dodson, "Predation, Body Size, and Composition of Plankton," *Science* 150, no. 3692 (1965): 28–35.
48. Stephen R. Carpenter et al., "Regulation of Lake Primary Productivity by Food Web Structure," *Ecology* 68, no. 6 (1987): 1863–76.
49. Stephen R. Carpenter, "Trophic Cascades in Lakes: Lessons and Prospects," in Terborgh and Estes, *Trophic Cascades*.
50. Ibid.
51. Mary Power, "Habitat Quality and the Distribution of Algae-Grazing Catfish in a Panamanian Stream," *Journal of Animal Ecology* 53 (1984): 357–74; Mary Power et al., "River-to-Watershed Subsidies in Old-Growth Conifer Forest," in *Food Webs at the Landscape Level*, ed. Gary A. Polis, Mary E. Power, and Gary R. Huxel (Chicago: University of Chicago Press, 2004), 217–40.
52. Robert J. Naiman, Robert E. Bilby, and Peter A. Bisson, "Riparian Ecology and Management in the Pacific Coastal Rain Forest," *BioScience* 50, no. 11 (2000): 996–1011.

53. Mary E. Power, "Effects of Fish in River Food Webs," *Science* 250, no. 4982 (1990): 811–14.
54. Ibid.
55. Mary E. Power, Michael S. Parker, and William E. Dietrich, "Seasonal Reassembly of a River Food Web: Floods, Droughts, and Impacts on Fish," *Ecological Monographs* 78, no. 2 (2008): 263–82.
56. Stuart A. Sandin, Sheila M. Walsh, and Jeremy B. C. Jackson, "Prey Release, Trophic Cascades, and Phase Shifts in Tropical Nearshore Ecosystems," in Terborgh and Estes, *Trophic Cascades*; Edward E. Demartini et al., "Differences in Fish-Assemblage Structure between Fished and Unfished Atolls in the Northern Line Islands, Central Pacific," *Marine Ecology Progress Series* 365 (2008): 199–215.
57. Edward O. Wilson, *The Diversity of Life* (Cambridge, MA: Belknap Press of Harvard University Press, 1992), 179.
58. Sandin, Walsh, and Jackson, "Prey Release, Trophic Cascades, and Phase Shifts."
59. Peter J. Mumby et al., "Trophic Cascade Facilitates Coral Recruitment in a Marine Reserve," *Proceedings of the National Academy of Sciences of the United States of America* 104, no. 20 (2007): 8362–67.
60. Sandin, Walsh, and Jackson, "Prey Release, Trophic Cascades, and Phase Shifts."

Chapter 4. Why the Earth Is Green: Terrestrial Cascades

1. Donald R. Strong, "Are Trophic Cascades All Wet? Differentiation and Donor Control in Speciose Ecosystems," *Ecology* 73, no. 3 (1992): 747–54; Bruce A. Menge, interview by Cristina Eisenberg, January 22, 2008, Oregon State University, Corvallis.
2. Oswald J. Schmitz, Peter A. Hamback, and Andrew P. Beckerman, "Trophic Cascades in Terrestrial Systems: A Review of the Effects of Carnivore Removals on Plants," *American Naturalist* 155, no. 2 (2000): 141–53.
3. Stewart A. Pickett and P. White, *The Ecology of Natural Disturbance and Patch Dynamics* (New York: Academic Press, 1986).
4. Richard B. Keigley, Michael R. Frisina, and Craig Fager, "A Method for Determining the Onset Year of Intense Browsing," *Journal of Range Management* 56, no. 1 (2003): 33–38.
5. John Terborgh, "Diversity: The Big Things That Run the World—a Sequel to E. O. Wilson," *Conservation Biology* 2, no. 4 (1988): 402–3.
6. John Terborgh, "The Green World Hypothesis Revisited," in *Large Carnivores and the Conservation of Biodiversity*, ed. Justina Ray et al. (Washington, DC: Island Press, 2005), 82–99.
7. John Terborgh et al., "Transitory States in Relaxing Ecosystems of Land-Bridge Islands," in *Tropical Forest Remnants: Ecology, Management, and Conservation of Frag-*

mented Communities, ed. William F. Laurance and Richard O. Bierregaard Jr. (Chicago: University of Chicago Press, 1997), 256–74.

8. John Terborgh et al., "Vegetation Dynamics of Predator-Free Land-Bridge Islands," *Journal of Ecology* 94, no. 2 (2006): 253–63.
9. William Stolzenburg, *Where the Wild Things Were: Life, Death, and Ecological Wreckage in a Land of Vanishing Predators* (New York: Bloomsbury, 2008), 95–97.
10. John Terborgh, interview by Cristina Eisenberg, February 13, 2008, Duke University, Durham, NC.
11. John Terborgh, *Requiem for Nature* (Washington, DC: Island Press, 2004), 120.
12. Jared Diamond, "Dammed Experiments!" *Science* 294, no. 5548 (2001): 1847–48.
13. Durward K. Allen, *Wolves of Minong: Their Vital Role in a Wild Community* (New York: Houghton Mifflin, 1979).
14. Charles C. Adams, "The Conservation of Predatory Mammals," *Journal of Mammalogy* 6, no. 2 (May 1925): 83–96; Adolph Murie, *The Moose of Isle Royale* (Ann Arbor: University of Michigan Press, 1934).
15. Rolf O. Peterson et al., "Temporal and Spatial Aspects of Predator-Prey Dynamics," *Alces* 39 (2003): 215–32.
16. Rolf Peterson, interview by Cristina Eisenberg, April 29, 2008, Michigan Technological University, Houghton.
17. Brian E. McLaren and Rolf O. Peterson, "Wolves, Moose, and Tree Rings on Isle Royale," *Science* 266, no. 5190 (1994): 1555–58.
18. Rolf Peterson, interview by Cristina Eisenberg, April 29, 2008, Michigan Technological University, Houghton.
19. Aldo Leopold, *A Sand County Almanac: And Sketches Here and There* (Oxford: Oxford University Press, 1949), 132.
20. George Desort, *Fortunate Wilderness: The Wolf and Moose Study of Isle Royale* (Michigan: Isinglass Pictures, 2008).
21. Christopher C. Wilmers et al., "Predator Disease Out-Break Modulates Top-Down, Bottom-Up, and Climatic Effects on Herbivore Population Dynamics," *Ecology Letters* 9 (2006): 383–89.
22. P. J. White, "2009 Annual Winter Trend Count of Northern Yellowstone Elk" (Yellowstone National Park, WY: National Park Service, Yellowstone Center for Resources, 2009); Douglas W. Smith et al., "Yellowstone Wolf Project: Annual Report, 2008," YCR-2009-03 (Yellowstone National Park, WY: National Park Service, Yellowstone Center for Resources, 2009).
23. Olaus J. Murie, notes on range conditions, O. J. Murie Papers, Teton Science School Archives, Murie Museum, Jackson, WY, 1926–54; Olaus J. Murie, *The Elk of North America* (Washington, DC: Wildlife Management Institute, 1951), 279.
24. Frederick H. Wagner, *Yellowstone's Destabilized Ecosystem* (New York: Oxford University Press, 2006), 3–11.

25. William H. Romme et al., "Aspen, Elk, and Fire in Northern Yellowstone Park," *Ecology* 76, no. 7 (1995): 2097–2106.
26. Charles E. Kay, "Are Ecosystems Structured from the Top-Down or Bottom-Up: A New Look at an Old Debate," *Wildlife Society Bulletin* 26, no. 3 (1998): 484–98.
27. Clifford A. White, Charles E. Olmsted, and Charles E. Kay, "Aspen, Elk, and Fire in the Rocky Mountain National Parks of North America," *Wildlife Society Bulletin* 26, no. 3 (1998): 449–62.
28. Michael E. Soulé and Reed Noss, "Rewilding and Biodiversity," *Wild Earth*, Fall 1998, 6.
29. Joel Berger et al., "A Mammalian Predator-Prey Imbalance: Grizzly Bear and Wolf Extinction Affect the Diversity of Avian Neotropical Migrants," *Ecological Applications* 11, no. 4 (2002): 947–60.
30. William J. Ripple and Eric J. Larsen, "Historic Aspen Recruitment, Elk, and Wolves in Northern Yellowstone National Park, USA," *Biological Conservation* 95, no. 3 (2000): 361–70.
31. Eric J. Larsen, interview by Cristina Eisenberg, November 20, 2008, University of Wisconsin, Stevens Point.
32. William J. Ripple, interview by Cristina Eisenberg, February 15, 2008, Oregon State University, Corvallis.
33. Edward R. Warren, "A Study of Beaver in the Yancey Region of Yellowstone National Park," *Roosevelt Wildlife Annals* 1 (1926): 1–191.
34. William J. Ripple et al., "Trophic Cascades among Wolves, Elk, and Aspen on Yellowstone National Park's Northern Range," *Biological Conservation* 102, no. 3 (2001): 227–34; Eric J. Larsen and William J. Ripple, "Aspen Age Structure in the Northern Yellowstone Ecosystem: USA," *Forest Ecology and Management* 179 (2003): 469–82.
35. Peterson et al., "Temporal and Spatial Aspects of Predator-Prey Dynamics."
36. Douglas Smith (director, Yellowstone Gray Wolf Restoration Project), interview by Cristina Eisenberg, November 4, 2008, Yellowstone National Park.
37. Robert L. Beschta, "Reduced Cottonwood Recruitment following Extirpation of Wolves in Yellowstone's Northern Range," *Ecology* 86, no. 2 (2005): 391–403.
38. Robert L. Beschta and William J. Ripple, "Recovering Riparian Plant Communities with Wolves in Northern Yellowstone, U.S.A.," *Restoration Ecology*, published online October 6, 2008, doi:10.1111/j.1526-100X.2008.00450.x; Douglas W. Smith and Daniel B. Stahler, "The Beavers of Yellowstone," *Yellowstone Science* 16, no. 3 (2008): 4–15.
39. White, "2009 Annual Winter Trend Count of Northern Yellowstone Elk"; John A. Vucetich, Douglas W. Smith, and Daniel R. Stahler, "Influence of Harvest, Climate, and Wolf Predation on Yellowstone Elk, 1961–2004," *Oikos* 111, no. 2 (2005): 259–70.
40. Douglas Smith (director, Yellowstone Gray Wolf Restoration Project), interview by Cristina Eisenberg, November 4, 2008, Yellowstone National Park.

41. William J. Ripple and Robert L. Beschta, "Wolf Reintroduction, Predation Risk, and Cottonwood Recovery in Yellowstone National Park," *Forest Ecology and Management* 184 (2003): 299–313.
42. Mark Hebblewhite, Evelyn H. Merrill, and Trent L. McDonald, "Spatial Decomposition of Predation Risk Using Resource Selection Functions: An Example in a Wolf-Elk Predator-Prey System," *Oikos* 111, no. 1 (2005): 101–11.
43. Matthew Kauffman et al., "Landscape Heterogeneity Shapes Predation in a Newly Restored Predator-Prey System," *Ecology Letters* 10, no. 8 (2007): 690–700.
44. Ibid.
45. Ibid.
46. Douglas Smith (director, Yellowstone Gray Wolf Restoration Project), interview by Cristina Eisenberg, November 4, 2008, Yellowstone National Park.
47. Valerius Geist, interview by Cristina Eisenberg, December 5, 2008, Houston, TX.
48. Mark Hebblewhite et al., "Human Activity Mediates a Trophic Cascade Caused by Wolves," *Ecology* 86, no. 8 (2005): 2135–44.
49. Anthony R. E. Sinclair, S. Mduma, and J. S. Brashares, "Patterns of Predation in a Diverse Predator-Prey System," *Nature* 425, no. 6955 (2003): 288–90.
50. Anthony R. E. Sinclair et al., "Long-Term Ecosystem Dynamics in the Serengeti: Lessons for Conservation," *Conservation Biology* 21, no. 5 (2007): 580–90.
51. Ibid.
52. Anthony R. E. Sinclair, "Mammal Population Regulation, Keystone Processes, and Ecosystem Dynamics," *Philosophical Transactions of the Royal Society of London B: Biological Sciences* 358, no. 1438 (2003): 1729–40.
53. Michael Soulé, personal communication, April 17, 2009, De Beque, Colorado.

Chapter 5. The Long View: Old-Growth Rain Forest Food Webs

1. Edward O. Wilson, *The Diversity of Life*, new ed. (New York: Norton, 1999), 196–98.
2. Jerry F. Franklin et al., "Ecological Characteristics of Old-Growth Douglas-Fir Forests," General Technical Report PNW-118 (Portland, OR: US Department of Agriculture, Forest Service, Pacific Northwest Research Station, 1981).
3. Margaret D. Lowman and H. Bruce Rinker, *Forest Canopies*, 2nd ed. (Amsterdam: Elsevier Academic Press, 2002), 3.
4. Daniel H. Janzen and Paul S. Martin, "Neotropical Anachronisms: The Fruits the Gomphotheres Ate," *Science* 215, no. 4528 (1982): 19–27; James Estes, interview by Cristina Eisenberg, March 31, 2008, Santa Cruz, CA.
5. Kent H. Redford, "The Empty Forest," *BioScience* 42, no. 6 (1992): 412–22.
6. John Terborgh et al., "Tree Recruitment in an Empty Forest," *Ecology* 89, no. 6 (2008): 1757–68.
7. Jerry Franklin, personal communication, February 9, 2009, Oregon State University, College of Forestry, Corvallis.

8. B. Thomas Parry, Henry J. Vaux, and Nicholas Dennis, "Changing Conceptions of Sustained-Yield Policy on the National Forests," *Journal of Forestry* 81, no. 3 (March 1983): 150–54; K. Norman Johnson and Frederick J. Swanson, "Historical Context of Old-Growth Forests in the Pacific Northwest: Policy, Practices, and Competing Worldviews," in *Old Growth in a New World: A Pacific Northwest Icon Reexamined*, ed. Thomas A. Spies and Sally L. Duncan (Washington, DC: Island Press, 2009), 12–28.
9. Frederick E. Clements, "Nature and Structure of the Climax," *Journal of Ecology* 24, no. 1 (1936): 252–84.
10. Donald Worster, *Nature's Economy: A History of Ecological Ideas*, 2nd ed. (Cambridge: Cambridge University Press, 1994), 300.
11. Aldo Leopold, "The Ecological Conscience," *Bulletin of the Garden Club of America*, September 1947, 45.
12. Ariel Lugo et al., "Long-Term Research at the USDA Forest Service's Experimental Forests and Ranges," *BioScience* 56, no. 1 (2006): 39–48.
13. Robert Michael Pyle, "The Long Haul," *Orion*, September–October 2004, 70–71.
14. Jon R. Luoma, *The Hidden Forest: The Biography of an Ecosystem* (Corvallis: Oregon State University Press, 2006), 69–84.
15. Pyle, "Long Haul," 70–71.
16. Chris Maser, Andrew W. Claridge, and James M. Trappe, *Trees, Truffles, and Beasts: How Forests Function* (New Brunswick, NJ: Rutgers University Press, 2008), 6–9, 52–57.
17. Ibid., 67–86.
18. Frederick J. Swanson et al., "Flood Disturbance in a Forested Mountain Landscape," *BioScience* 48, no. 9 (1998): 681–89.
19. Luoma, *Hidden Forest*, 141–45.
20. Franklin et al., "Ecological Characteristics of Old-Growth Douglas-Fir Forests."
21. Stanley V. Gregory et al., "An Ecosystem Perspective of Riparian Zones," *BioScience* 41, no. 8 (1991): 540–53.
22. Richard Cannings and Jared Hobbs, *Spotted Owls: Shadows in an Old-Growth Forest* (Vancouver, British Columbia: Greystone Books, 2007), 31.
23. Johnson and Swanson, "Historical Context of Old-Growth Forests in the Pacific Northwest."
24. Jerry F. Franklin, "Scientists in Wonderland," *BioScience* 45 (1995): S74–S78.
25. Jack Ward Thomas et al., "The Northwest Forest Plan: Origins, Components, Implementation Experience, and Suggestions for Change," *Conservation Biology* 20, no. 2 (2006): 277–87.
26. Ibid.
27. Frederick J. Swanson, personal communication, December 6, 2009.
28. Oregon Department of Fish and Wildlife, "Wolf Pack with Pups Confirmed in

Northeastern Oregon," July 21, 2008, http://www.dfw.state.or.us/news/2008/july/072108b.asp, accessed April 30, 2009.

29. Vernon Bailey, *North American Fauna: Mammals of Oregon* (Washington, DC: US Department of Agriculture, Bureau of Biological Survey, 1936), 272–75; Oregon Department of Fish and Wildlife, "Oregon Elk Management Plan," February 2003, http://www.dfw.state.or.us/wildlife/management_plans, accessed April 29, 2009.
30. Frederick J. Swanson, personal communication, Victoria, British Columbia, June 5, 2009; Jerry Franklin, e-mail to Cristina Eisenberg, June 12, 2009.
31. Frederick J. Swanson, Charles Goodrich, and Kathleen Dean Moore, "Bridging Boundaries: Scientists, Creative Writers, and the Long View of the Forest," *Frontiers in Ecology and the Environment* 6, no. 9 (2008): 499–504.
32. Alice P. Meyers and Nancy Fredricks, "Thornton T. Munger Research Natural Area Management Plan, Gifford Pinchot National Forest Wind River Ranger District" (Portland, OR: US Department of Agriculture, Forest Service, Pacific Northwest Research Station, 1993); Margaret Herring and Sarah Greene, *Forest of Time: A Century of Science at Wind River Experimental Forest* (Corvallis: Oregon State University, 2007).
33. David C. Shaw et al., "Ecological Setting of the Wind River Old-Growth Forest," *Ecosystems* 7, no. 5 (2004): 427–39.
34. David Shaw and Ken Bible, "An Overview of Forest Canopy Ecosystem Functions with Reference to Urban and Riparian Systems," *Northwest Science* 70 (1996): 1–6.
35. Franklin et al., "Ecological Characteristics of Old-Growth Douglas-Fir Forests," 56.
36. Andrew B. Carey, "Interactions of Northwest Forest Canopies and Arboreal Mammals," *Northwest Science* 70 (1996): 72–78; David C. Shaw and Catherine Flick, "Are Resident Songbirds Stratified within the Canopy of a Coniferous Old-Growth Forest?" *Selbyana* 20, no. 2 (1999): 324–31.
37. Lowman and Rinker, *Forest Canopies*, 3–11; Bruce McCune et al., "Vertical Profile of Epiphytes in a Pacific Northwest Old-Growth Forest," *Northwest Science* 71, no. 2 (1997): 145–52.
38. Ray Bradbury, *A Sound of Thunder and Other Stories* (New York: Harper Perennial, 2005).
39. The Xerces Society for Invertebrate Conservation, "Hairstreaks: Johnson's hairstreak (*Callophrys johnsoni*)," http://www.xerces.org/johnsons-hairstreak/, accessed May 26, 2009; Robert Michael Pyle, *The Butterflies of Cascadia: A Field Guide to All the Species of Washington, Oregon, and Surrounding Territories* (Seattle, WA: Seattle Audubon Society, 2002), 210.
40. C. David McIntire, J. D. Hall, and Duane Lee Higley, "Report of the Aquatic Modeling Group: Round One," Internal Report (US International Biological Program, Coniferous Forest Biome) no. 29 (Seattle: University of Washington, 1972).

41. Steven M. Wondzell, "Floods, Channel Change, and the Hyporheic Zone," *Water Resources Research* 35, no. 2 (1999): 555–67; Mary Power et al., "River-to-Watershed Subsidies in Old-Growth Conifer Forest," in *Food Webs at the Landscape Level*, ed. Gary A. Polis, Mary E. Power, and Gary R. Huxel (Chicago: University of Chicago Press, 2004), 217–40.
42. Julia A. Jones and David A. Post, "Seasonal and Successional Streamflow Response to Forest Cutting and Regrowth in the Northwest and Eastern United States," *Water Resources Research* 40 (2004): W05203, doi:10.1029/2003WR002952.
43. Robert J. Naiman and Kevin H. Rogers, "Large Animals and System-Level Characteristics in River Corridors: Implications for River Management," *BioScience* 47, no. 8 (1997): 521–29; Robert J. Naiman et al., "A Process-Based View of Floodplain Forest Patterns in Coastal River Valleys of the Pacific Northwest," *Ecosystems*, published online November 14, 2009, doi:10.1007/s10021-009-9298-5; Robert J. Naiman, Robert E. Bilby, and Peter A. Bisson, "Riparian Ecology and Management in the Pacific Coastal Rain Forest," *BioScience* 50, no. 11 (2000): 996–1011.
44. Mark E. Harmon and Jerry F. Franklin, "Age Distribution of Western Hemlock and Its Relation to Roosevelt Elk Populations in the South Fork Hoh River Valley, Washington," *Northwest Science* 57, no. 4 (1983): 249–55; Olaus J. Murie, "Report on the Elk of the Olympic Peninsula" (Jackson, WY: US Biological Survey, 1935).
45. Edward G. Schreiner et al., "Understory Patch Dynamics and Ungulate Herbivory in Old-Growth Forests of Olympic National Park, Washington," *Canadian Journal of Forest Research* 26 (1996): 255–65; Andrea Woodward et al., "Ungulate-Forest Relationships in Olympic National Park: Retrospective Exclosure Studies," *Northwest Science* 68, no. 2 (1994): 97–102; Jerry Franklin, e-mail to Cristina Eisenberg, May 7, 2009.
46. Jerry Franklin, e-mail to Cristina Eisenberg, May 7, 2009.
47. Robert L. Beschta and William J. Ripple, "Wolves, Trophic Cascades, and Rivers in the Olympic National Park, USA," *Ecohydrology* 1 (2008): 118–30.
48. Naiman, Bilby, and Bisson, "Riparian Ecology and Management"; Jim E. O'Connor, Myrtle A. Jones, and Tana L. Haluska, "Flood Plain and Channel Dynamics of the Quinault and Queets Rivers, Washington, USA," *Geomorphology* 51 (2003): 31–59; Robert Naiman, e-mail to Cristina Eisenberg, May 11, 2009; Frederick J. Swanson, e-mail to Cristina Eisenberg, May 7, 2009.
49. William S. Alverson, Donald M. Waller, and Stephen L. Solheim, "Forests Too Deer: Edge Effects in Northern Wisconsin," *Conservation Biology* 2, no. 4 (1988): 348–58.
50. Jerry Franklin, e-mail to Cristina Eisenberg, May 7, 2009.
51. Ken Bible, personal communication, March 26, 2009, Wind River Experimental Forest, Stabler, WA; Andrew J. Larson and Robert T. Paine, "Ungulate Herbivory: Indirect Effects Cascade into the Treetops," *Proceedings of the National Academy of Sciences of the United States of America* 104, no. 1 (2007): 5–6.

Chapter 6. All Our Relations: Trophic Cascades and the Diversity of Life

1. Michael Soulé, "What Is Conservation Biology?" in *Environmental Policy and Biodiversity*, ed. R. Edward Grumbine (Washington, DC: Island Press, 1994), 35.
2. James Maclaurin and Kim Sterelny, *What Is Biodiversity?* (Chicago: University of Chicago Press, 2008), 2.
3. Reed F. Noss and Allen Y. Cooperrider, *Saving Nature's Legacy: Protecting and Restoring Biodiversity* (Washington, DC: Island Press, 1994), 5.
4. Edward O. Wilson, *The Diversity of Life*, new ed. (New York: Norton, 1999), 38; Robert T. Paine, interview by Cristina Eisenberg, January 31, 2008, University of Washington, Seattle.
5. R. Edward Grumbine, *Ghost Bears: Exploring the Biodiversity Crisis* (Washington, DC: Island Press, 1992), 22–25.
6. Wilson, *Diversity of Life*, 164–70; quotation, 364.
7. Edward O. Wilson, *Naturalist* (Washington, DC: Island Press, 1994), 5–15.
8. Robert MacArthur and Edward O. Wilson, *The Theory of Island Biogeography* (Princeton, NJ: Princeton University Press, 1967).
9. David Quammen, *The Song of the Dodo: Island Biogeography in an Age of Extinctions* (New York: Scribner, 1996), 141–45.
10. Wilson, *Naturalist*, 209–17; Quammen, *Song of the Dodo*, 418–23; Daniel Simberloff, "Experimental Zoogeography of Islands: The Colonization of Empty Islands," *Ecology* 50, no. 2 (1969): 278–95.
11. Daniel S. Simberloff and Lawrence G. Abele, "Island Biogeography Theory and Conservation Practice," *Science* 191, no. 4224 (1976): 285–86.
12. Wilson, *Diversity of Life*, 280.
13. James A. Estes, David O. Duggins, and Galen B. Rathbun, "The Ecology of Extinctions in Kelp Forest Communities," *Conservation Biology* 4, no. 3 (1989): 252–64.
14. David S. Wilcove, *The Condor's Shadow* (New York: Freeman, 1999), 7–8.
15. Reed F. Noss, Edward T. LaRoe III, and J. Michael Scott, *Endangered Ecosystems of the United States: A Preliminary Assessment of Loss and Degradation* (Washington, DC: US Geological Survey, 1995); John Terborgh, *Requiem for Nature* (Washington, DC: Island Press, 1999), 16–17.
16. Hal Salwasser, "In Search of an Ecosystem Approach to Endangered Species Conservation," in *Balancing on the Brink of Extinction: The Endangered Species Act and Lessons for the Future*, ed. Kathryn A. Kohm (Washington, DC: Island Press, 1991), 247–65.
17. G. Evelyn Hutchinson, "Homage to Santa Rosalia, or Why Are There So Many Kinds of Animals?" *American Naturalist* 93, no. 870 (1959): 145–59.
18. Wilson, *Diversity of Life*, 308.
19. Harold A. Mooney et al., "What We Have Learned about the Ecosystem Functioning

of Biodiversity," in *Functional Roles of Biodiversity: A Global Perspective*, ed. Harold A. Mooney et al. (New York: Wiley, 1996), 476–84.

20. Jerry Melillo and Osvaldo Sala, "Ecosystem Services," in *Sustaining Life: How Human Health Depends on Biodiversity*, ed. Eric Chivian and Aaron Bernstein (Oxford: Oxford University Press, 2008), 74–115; Peter B. Reich et al., "Plant Diversity Enhances Ecosystem Responses to Elevated CO_2 and Nitrogen Deposition," *Nature* 410 (2001): 809–10.
21. Wilson, *Diversity of Life*, 287.
22. Ibid., 177–79.
23. Terborgh, *Requiem for Nature*, 144.
24. As quoted in Dave Foreman, *Rewilding North America: A Vision for Conservation in the 21st Century* (Washington, DC: Island Press, 2004), 205.
25. Hal Salwasser (dean, Oregon State University College of Forestry; director, Pacific Forest Laboratory), interview by Cristina Eisenberg, January 24, 2008, Oregon State University, Corvallis.
26. Robert T. Paine, "A Conversation on Refining the Concept of Keystone Species," *Conservation Biology* 9, no. 4 (1995): 962–64; Wilson, *Diversity of Life*, 165.
27. Robert Steneck and Enric Sala, "Large Marine Carnivores: Trophic Cascades and Top-Down Controls in Coastal Ecosystems Past and Present," in *Large Carnivores and the Conservation of Biodiversity*, ed. Justina Ray et al. (Washington, DC: Island Press, 2005), 112–16.
28. Lee A. Dyer and Phyllis D. Coley, "Tritrophic Interactions in Tropical versus Temperate Communities," in *Multitrophic Level Interactions*, ed. Teja Tscharntke and Bradford A. Hawkins (Cambridge: Cambridge University Press, 2002), 67–84; Niles Eldredge, *Life in the Balance: Humanity and the Biodiversity Crisis* (Princeton, NJ: Princeton University Press, 1998), 113–15.
29. Donald R. Strong, "Are Trophic Cascades All Wet? Differentiation and Donor Control in Speciose Ecosystems," *Ecology* 73, no. 3 (1992): 747–54.
30. Wilson, *Diversity of Life*, 150.
31. Justina Ray, "Large Carnivorous Animals as Tools for Conserving Biodiversity: Assumptions and Uncertainties," in Ray, *Large Carnivores and the Conservation of Biodiversity*, 34–56.

Chapter 7. Creating Landscapes of Hope: Trophic Cascades and Ecological Restoration

1. Marissa A. Ahlering and John Faaborg, "Avian Habitat Management Meets Conspecific Attraction: If You Build It, Will They Come?" *Auk* 123, no. 2 (2006): 301–12.
2. Society for Ecological Restoration International, Science and Policy Working Group, "The SER International Primer on Ecological Restoration" (Tucson, AZ: So-

ciety for Ecological Restoration International, 2004), http://www.ser.org, accessed December 2, 2008.

3. M. Jake Vander Zanden, Julian D. Olden, and Claudio Gratton, "Food-Web Approaches in Restoration Ecology," in *Foundations of Restoration Ecology*, ed. Donald A. Falk, Margaret A. Palmer, and Joy B. Zedler (Washington, DC: Island Press, 2006), 165–89.
4. Society for Ecological Restoration International, "SER International Primer on Ecological Restoration."
5. Margaret A. Palmer, Richard F. Ambrose, and N. LeRoy Poff, "Ecological Theory and Community Restoration Ecology," *Restoration Ecology* 5, no. 4 (1997): 291–300.
6. Steven G. Whisenant, *Repairing Damaged Wildlands: A Process-Oriented, Landscape-Scale Approach* (Cambridge: Cambridge University Press, 1999).
7. Margaret Palmer et al., "Ecology for a Crowded Planet," *Science* 304, no. 5675 (2004): 1251–52.
8. James F. Kitchell, ed., *Food Web Management: A Case Study of Lake Mendota* (New York: Springer, 1992).
9. P. S. Lake et al., "Global Change and the Biodiversity of Freshwater Ecosystems: Impacts on Linkages between Above-Sediment and Sediment Biota," *BioScience* 50, no. 12 (2000): 1099–1106.
10. James F. Kitchell, "The Rationale and Goals for Food Web Management in Lake Mendota," in Kitchell, *Food Web Management*, 1–4.
11. Richard C. Lathrop et al., "Stocking Piscivores to Improve Fishing and Water Clarity: A Synthesis of the Lake Mendota Biomanipulation Project," *Freshwater Biology* 47 (2002): 2410–24.
12. James F. Kitchell and Stephen R. Carpenter, "Accomplishments and New Directions of Food Web Management in Lake Mendota," in Kitchell, *Food Web Management*, 539–40.
13. Brian Reed Silliman and Jay C. Zieman, "Top-Down Control of *Spartina alterniflora* Production by Periwinkle Grazing in a Virginia Salt Marsh," *Ecology* 82, no. 10 (2001): 2830–45.
14. Brian Reed Silliman and Mark D. Bertness, "A Trophic Cascade Regulates Salt Marsh Primary Production," *Proceedings of the National Academy of Sciences of the United States of America* 99, no. 16 (2002): 10500–5.
15. William J. Ripple, interview by Cristina Eisenberg, February 15, 2008, Oregon State University, Corvallis.
16. Robert L. Beschta, interview by Cristina Eisenberg, January 14, 2008, Oregon State University, Corvallis.
17. Robert L. Beschta and William J. Ripple, "Recovering Riparian Plant Communities with Wolves in Northern Yellowstone, U.S.A.," *Restoration Ecology*, published online October 6, 2008, doi:10.1111/j.1526-100X.2008.00450.x.

18. William J. Ripple and Robert L. Beschta, "Linking a Cougar Decline, Trophic Cascade, and Catastrophic Regime Shift in Zion National Park," *Biological Conservation* 133 (2006): 397–408.
19. William J. Ripple and Robert L. Beschta, "Trophic Cascades Involving Cougar, Mule Deer, and Black Oaks in Yosemite National Park," *Biological Conservation* 141 (2008): 1249–56.
20. Rolf Peterson, interview by Cristina Eisenberg, April 29, 2008, Michigan Technological University, Houghton; Douglas Smith (director, Yellowstone Gray Wolf Restoration Project), interview by Cristina Eisenberg, November 4, 2008, Yellowstone National Park.
21. William J. Ripple and R. L. Beschta, "Refugia from Browsing as Reference Sites for Restoration Planning," *Western North American Naturalist* 65, no. 2 (2005): 269–73.
22. Mark A. Davis and Lawrence B. Slobodkin, "The Science and Values of Restoration Ecology," *Restoration Ecology* 12 (2004): 1–3.
23. John Rappold, interview by Cristina Eisenberg, November 3, 2008, Dupuyer, MT.
24. John Russell and Charlie Russell, interview by Cristina Eisenberg, October 14, 2008, the Hawk's Nest, Russell Ranch, Waterton, Alberta.
25. Paul Vahldiek, interview by Cristina Eisenberg, October 14, 2008, the Hawk's Nest, Russell Ranch, Waterton, Alberta.
26. Ibid.
27. Eric W. Sanderson et al., "The Ecological Future of the North American Bison: Conceiving Long-Term, Large-Scale Conservation of Wildlife," *Conservation Biology* 22, no. 2 (2008): 252–66; Carron A. Meany and Dirk van Vuren, "Recent Distribution of Bison in Colorado West of the Great Plains," *Proceedings of the Denver Museum of Natural History* 3, no. 4 (1993): 1–10.
28. Roger Creasey, interview by Cristina Eisenberg, October 14, 2008, the Hawk's Nest, Russell Ranch, Waterton, Alberta.

Chapter 8. Finding Common Ground: Trophic Cascades and Ecosystem Management

1. Dave Foreman, *Rewilding North America: A Vision for Conservation in the 21st Century* (Washington, DC: Island Press, 2004), 23.
2. John Terborgh and Michael E. Soulé, "Why We Need Mega-Reserves—and How to Design Them," in *Continental Conservation: Scientific Foundations of Regional Reserve Networks*, ed. Michael E. Soulé and John Terborgh (Washington, DC: Island Press, 1999), 199–209; quotation, 202.
3. Brian Miller et al., "The Importance of Large Carnivores to Healthy Ecosystems," *Endangered Species Update* 18, no. 5 (2001): 202–10.
4. Dave Foreman (director, The Rewilding Institute), interview by Cristina Eisenberg, March 8, 2008, University of Oregon, Eugene.

5. Foreman, *Rewilding North America*, 127–29.
6. Michael E. Soulé and Reed Noss, "Rewilding and Biodiversity," *Wild Earth*, Fall 1998, 19–28.
7. Dave Foreman (director, The Rewilding Institute), interview by Cristina Eisenberg, March 8, 2008, University of Oregon, Eugene; Foreman, *Rewilding North America*, 138–40. The names of the wildways given in this chapter differ slightly from those in Foreman's book, reflecting updates at http://www.wildlandsproject.org as of December 2009.
8. Aldo Leopold, *Round River* (Oxford: Oxford University Press, 1971), 197.
9. Soulé and Noss, "Rewilding and Biodiversity."
10. US Department of the Interior, "Final Rule to Establish a Gray Wolf–Northern Rocky Mountain Distinct Population Segment and Remove from the Federal List of Threatened and Endangered Species," *Federal Register* (February 21, 2008): 68/62, 15804–82, 50 CFR Part 17.
11. US Census Bureau, "U.S. and World Population Clocks," http://www.census.gov/main/www/popclock.html, accessed February 22, 2009; J. E. Cohen, "Human Population: The Next Half Century," *Science* 307, no. 5648 (2003): 1172–75.
12. Anthony R. E. Sinclair et al., "Testing Hypotheses of Trophic Level Interactions: A Boreal Forest Ecosystem," *Oikos* 89, no. 2 (2000): 313–28.
13. Timothy W. Clark, Paul C. Paquet, and A. P. Curlee, "Large Carnivore Conservation in the Rocky Mountains of the United States and Canada," *Conservation Biology* 10, no. 4 (1996): 936–39.
14. Aldo Leopold, *A Sand County Almanac: And Sketches Here and There* (Oxford: Oxford University Press, 1989), 224–25.
15. Aldo Leopold, "A Biotic View of the Land," *Journal of Forestry* 37 (1939): 729.
16. R. Edward Grumbine, *Ghost Bears: Exploring the Biodiversity Crisis* (Washington, DC: Island Press, 1992).
17. Hal Salwasser (dean, Oregon State University College of Forestry; director, Pacific Forest Laboratory), interview by Cristina Eisenberg, January 24, 2008, Oregon State University, Corvallis.
18. Gifford Pinchot, *Breaking New Ground* (New York: Harcourt Brace, 1947), 271.
19. Robert T. Lackey, "Seven Pillars of Ecosystem Management," *Landscape and Urban Planning* 40 (1998): 21–30; Hal Salwasser, "Ecosystem Management: A New Perspective for National Forests and Grasslands," in *Ecosystem Management: Adaptive Strategies for Natural Resources Organizations in the Twenty-first Century*, ed. William Burch (Levittown, PA: Taylor and Francis, 1999), 85–96.
20. Hal Salwasser, "Ecosystem Management: Can It Sustain Diversity and Productivity?" *Journal of Forestry* 92, no. 8 (1994): 6–10.
21. R. Edward Grumbine, "What Is Ecosystem Management?" *Conservation Biology* 8, no. 1 (1994): 31.
22. Lackey, "Seven Pillars of Ecosystem Management," 23.

23. Jerry F. Franklin, "Ecosystem Management: An Overview," in *Ecosystem Management: Applications for Sustainable Forest and Wildlife Resources*, ed. Mark S. Boyce and Alan Haney (New Haven, CT: Yale University Press, 1996).
24. Frederick E. Clements, "Nature and Structure of the Climax," *Journal of Ecology* 24, no. 1 (1936): 252–84; Charles Elton, *Animal Ecology*, 2nd ed. (Chicago: University of Chicago Press, 2001).
25. K. Norman Johnson, interview by Cristina Eisenberg, April 4, 2008, Oregon State University, Corvallis.
26. Norman L. Christensen et al., "The Report of the Ecological Society of America Committee on the Scientific Basis for Ecosystem Management," *Ecological Applications* 6, no. 3 (1996): 665–91.
27. Hal Salwasser (dean, Oregon State University College of Forestry; director, Pacific Forest Laboratory), interview by Cristina Eisenberg, January 24, 2008, Oregon State University, Corvallis.
28. Jerry Franklin, interview by Cristina Eisenberg, April 3, 2008, Corvallis, OR.
29. Lackey, "Seven Pillars of Ecosystem Management."
30. Frederick J. Swanson, interview by Cristina Eisenberg, January 8, 2009, Oregon State University, Corvallis.
31. Christensen et al., "Scientific Basis for Ecosystem Management."
32. Montana Fish, Wildlife, and Parks, "Final Environmental Impact Statement, Montana Gray Wolf Conservation and Management Plan" (Helena, MT: Montana Fish, Wildlife, and Parks, 2003).
33. Michael E. Soulé et al., "Strongly Interacting Species: Conservation Policy, Management, and Ethics," *BioScience* 55, no. 2 (2005): 168–76; James A. Estes, M. Tim Tinker, and James L. Bodkin, "Using Trophic Cascades to Develop Ecosystem-Based Recovery Criteria for Depleted Species: Sea Otters and Kelp Forests in the Aleutian Archipelago," *Conservation Biology* (forthcoming); Cristina Eisenberg et al., "Ecologically Effective Populations and Trophic Cascades: The Potential for Wolf Recovery to Shift Conservation Policy," *BioScience* (in preparation).
34. *Endangered Species Act of 1973*, Public Law 93-205, *US Statutes at Large* 87 (1988): 884, codified at *US Code* 16, §§ 1531–44, § 2 (4) c; Eisenberg et al., "Ecologically Effective Populations and Trophic Cascades."
35. *National Forest Management Act of 1976*, Public Law 94-588, *US Statutes at Large* 90 (October 22, 1976): 2949, § 6 (g) (3) (B).
36. Hal Salwasser, "Top Predators in National Forests," lecture delivered February 13, 2009, Oregon State University, Corvallis.
37. Estes, Tinker, and Bodkin, "Using Trophic Cascades to Develop Ecosystem-Based Recovery Criteria."
38. Eisenberg et al., "Ecologically Effective Populations and Trophic Cascades."
39. C. S. Holling, ed., *Adaptive Environmental Assessment and Management* (New York: Wiley, 1978); Roger E. Latham et al., "Managing White-Tailed Deer in Forest Habi-

tat from an Ecosystem Perspective: Pennsylvania Case Study" (Harrisburg, PA: Audubon Pennsylvania and Pennsylvania Habitat Alliance, 2005), 15–16.

40. K. Norman Johnson et al., *Forest Management: Conservation Strategies for a Changing and Uncertain World* (New York: McGraw-Hill, forthcoming).
41. Ibid.
42. Frederick Swanson et al., "Guide to Effective Research-Management Collaboration at Long-Term Environmental Research Sites" (Corvallis, OR: US Department of Agriculture, Forest Service, Pacific Northwest Research Station, forthcoming); Bernard T. Borman, Richard W. Haynes, and Jon R. Martin, "Adaptive Management of Forest Ecosystems: Did Some Rubber Hit the Road?" *BioScience* 57, no. 2 (2007): 186–91.
43. Frederick J. Swanson, interview by Cristina Eisenberg, February 17, 2009, Oregon State University, Corvallis.
44. Joshua Halofsky, e-mail to Cristina Eisenberg, December 5, 2008.

Epilogue. Lessons from 763

1. This number is fictitious to protect her identity.
2. Michael Soulé, "Conservation Relevance of Ecological Cascades," in *Trophic Cascades*, ed. John Terborgh and James A. Estes (Washington, DC: Island Press, forthcoming).
3. John Terborgh and James A. Estes, "Conclusion: Our Trophically Downgraded Planet," in Terborgh and Estes, *Trophic Cascades*.

Glossary

abiotic A term that describes nonliving elements that influence the growth, composition, and structure of a biotic community (e.g., soil, climate, topography).

adaptive management A science-based approach that acknowledges uncertainty and views policy and management decisions as testable hypotheses, to be revised using new information.

autotrophs Organisms that form the base of the food chain, also referred to as *primary producers*. On land they are plants; in aquatic environments they are algae.

background extinction The normal rate of extinction, usually one to five species per year.

benthic Pertaining to the bottom of a body of water.

biodiversity The variety of living organisms at all levels of organization, including genetic, species, and higher taxonomic levels; the variety of *habitats* and ecosystems, as well as the processes occurring therein.

biomass The total mass of living matter within a given area.

biota Plants and animals, which include all the organisms in an ecosystem.

bivalves A large group of freshwater and saltwater organisms that have a pair of shells (valves).

bottom-up control Regulation of *food web* components in an ecosystem either by the actions of primary producers (plants) or by limits on input of nutrients.

browsing Feeding on the leaves, branches, or shoots of *woody plants*, such as shrubs and trees.

carrying capacity In reference to animal populations, the optimum population level at which *habitat* is not limiting, or the steady-state density that a species can achieve in a particular habitat to support it sustainably, referred to as *K*. There are also *social* and *economic* carrying capacities, which are the density a species can achieve in places subject to multiple human land uses.

commensalism An interaction between two species in which one experiences a positive effect and the other experiences no effect.

community ecology The study of the structure of communities and how it varies in time and space in response to physical and biotic factors.

competition A nontrophic interaction between two species that is negative for both.

conservation biology An interdisciplinary approach to protecting and managing *biodiversity* that uses genetics, *ecology*, wildlife biology, anthropology, sociology, philosophy, and economics, as well as other fields, such as the creative arts and communications.

conspecific Belonging to the same species.

consumers A term used in *community ecology* for herbivores.

control A group of experimental subjects or an area not exposed to the treatment being applied or investigated, to be used for comparison.

critical habitat According to the Endangered Species Act, the ecosystems upon which endangered and threatened species depend.

cyanobacteria Photosynthetic bacteria, also called blue-green algae, that can perform *nitrogen fixation*.

density-dependent factors Effects on a population that change in relative intensity as population density changes, such as factors that affect the birthrate or mortality of a species. These factors typically include predation.

detritivore A species that obtains nutrients by consuming decomposing matter, such as rotting vegetation or animal flesh.

disturbance Any event that disrupts an ecosystem and changes nutrient flow or the structure and condition of the physical environment.

dominant species Species that owe their influences to their high abundance and that account for most of the *biomass* in a community, and thus are primary components of community structure.

ecological extinction The reduction of a species to such low abundance that although still present in the community, it no longer interacts significantly with other species.

ecological integrity The ability of an ecosystem to self-correct and return to the state normal for that system after a disturbance.

ecologically effective population A population of a *keystone species* of sufficient density and distribution to cause a trophic cascade.

ecological release *Habitat* expansion or density increase of a species in the absence of one or more competing species.

ecological restoration Assistance in the recovery of an ecosystem that has been degraded, damaged, or destroyed. The goal of this process is to emulate the structure, function, diversity, and dynamics of the specified ecosystem, not necessarily to return to the past.

ecology The study of the distribution and abundance of living organisms and their interactions with one another and with their environment.

ecoregion A relatively large area of land or water that contains a geographically distinct assemblage of natural communities.

ecotone The border between two different ecosystems, or *habitat* types, such as where field and forest meet.

empirical Derived from experiment and observation rather than theory; based on data.

endemic A term describing any localized process or pattern but usually applied to a highly localized or restrictive geographic distribution of a species.

eutrophication The slow, natural aging process during which a lake, estuary, or bay evolves into a bog or marsh and eventually disappears. During the later stages of eutrophication the body of water is choked by abundant plant life as a result of high levels of nutritive compounds, such as nitrogen and phosphorus.

exploitation ecosystems hypothesis This hypothesis, also referred to as EEH, suggests that primary productivity in an environment determines its number of *trophic* levels, with less productive systems, such as arctic tundra, having one to two trophic levels, and more productive systems, such as temperate forests, having three or more trophic levels.

extinction debt A situation wherein decades or centuries after a *habitat* perturbation, extinction related to it may still be taking place; a protracted extinction.

facilitation Improvement of resource availability for one species as a result of prior use of the resource by a second species.

feedback A phenomenon wherein a system's output modifies input to the system, thus becoming a self-sustaining situation. Prices play this role in market systems.

flagship species A charismatic species with broad popular appeal.

focal species A species whose requirements for survival represent factors important to ecosystem function. Scientists and managers pay attention to focal species because budgetary limitations prevent them from paying attention to all species, and focal species can address broad issues.

foliar biomass The total mass of the leaves of a tree, shrub, or *herbaceous plant.*

food web *Trophic* structures based on description or observation of linkages in a community (see *interaction web*).

functional groups Categories of organisms that play key roles in an ecosystem, such as primary producers, herbivores, carnivores, decomposers, nitrogen fixers, and pollinators.

grazing Feeding on *herbaceous plants* in a field or pasture.

green world hypothesis A hypothesis that suggests that the world is green because top predators control their herbivore prey, via predation, thereby having an indirect effect on vegetation, enabling it to grow.

habitat The dwelling place of an organism or community, which provides the necessary conditions for its existence.

habitat fragmentation The disruption of extensive habitats into isolated and small patches; often applied to forested habitats that have been fragmented by agricultural development or logging.

hectare A metric unit equivalent to 2.471 acres.

herbaceous plants Plants whose leaves and stems die down to soil level at the end of the growing season.

herbivory A form of predation in which an organism, known as an *herbivore*, consumes plants, also referred to as *autotrophs*.

heterogeneous Consisting of dissimilar elements; diverse. In *ecology* this term is used to refer to landscape, ecosystem, or community structure.

heuristic Serving to indicate or point out; stimulating interest as a means of furthering investigation; a teaching method that encourages students to make discoveries for themselves.

indicator species A species indicative of overall ecosystem function. Tied to a specific biotic community, it occurs only under a particular set of circumstances. Wildlife managers often use indicator species as a shortcut to monitoring a whole ecosystem.

interaction web *Trophic* structure based on experimental identification of strong links in a community. This represents a subset of species that through their interactions and responses to *abiotic* factors make up the dynamic core of *food webs* or communities. They can include *keystone species*, *dominant species*, and other strong interactors. Interaction webs are sometimes also referred to as *functional webs.*

irruption A sudden, explosive increase in population numbers, which if unchecked will exceed the capacity of the resource base to sustain the population. It often involves an exponential increase in population.

island biogeography The geographic distribution of plants and animals on islands.

keystone species A species whose impacts on its community or ecosystem are much larger than would be expected from its abundance. A keystone species is a carnivore that consumes and controls a particular prey species, which then competes with and excludes other species in its trophic class.

land-bridge islands Areas that presently are island *habitats* but formerly were connected to the mainland during periods of lower water levels. Land-bridge islands tend to lose species over time in a process called *relaxation.*

landscape matrix The intervening area among a set of *habitat* fragments. Also the spatial array of habitats across a landscape.

lentic Pertaining to or living in still water.

limnology The scientific study of bodies of freshwater, such as lakes and ponds.

littoral Pertaining to the shores of a lake, sea, or ocean.

littoral zone Another name for the *rocky intertidal zone*, a narrow zone between the coast and the ocean revealed at low tide, characterized by breaking waves and an environment exposed to the air for part of the day. This can also refer to the zone between the shore and a lake.

megafauna Species that weigh more than 100 pounds as adults.

megaherbivore Species, such as elephants, usually too large as adults to be killed by predators.

mesopredators Medium-sized predators, such as coyotes, raccoons, and foxes, which often increase in abundance when larger predators are eliminated.

meta-analysis The process of analyzing research in a specific area of inquiry by comparing and combining results from previous separate but related studies.

metapopulation A large population of animals made up of smaller populations, which interbreed and move freely between each other's ranges.

multifactorial experiment Research that involves creating several plots or treatment areas in which the researcher manipulates certain variables and then measures the effect of this manipulation. This classic form of research design enables scientists to make inferences about cause and effect.

mutualism An interspecific relationship in which both organisms benefit, frequently a relationship of complete dependence. Pollination is an example of this.

natural selection A theory created by Charles Darwin proposing that only the organisms best adapted to their environment survive, and that this process shapes traits in all organisms (for example, an antelope's fleet limbs).

net primary production Also called NPP, this refers to energy flow in ecosystems and is a measure of *biomass* or the ability of things to grow. Its components are energy flow via sunlight, moisture, and photosynthesis.

niche The position or role of a species in an ecosystem. It is popularly defined as how a species makes its living.

nitrogen fixation Part of the nitrogen cycle, which involves the process of various bacteria incorporating atmospheric nitrogen so that it can be usable as a nutrient. In aquatic communities *cyanobacteria* are primary nitrogen fixers.

old fields Impoverished lands, often characterized by poor soil, that have been farmed or grazed and subsequently abandoned. They represent an intermediate state of ecological plant *succession*, one dominated by grasses, forbs, and encroaching woody species.

order of magnitude A tenfold change in number that represents one exponential level plus or minus.

overexploitation The consumptive use of a natural resource beyond its capacity to replenish what has been taken.

Paleolithic Relating to the Stone Age, a period that began 2.5–2 million years ago and ended around 10,000 years ago.

paradigm An established pattern of thinking. Often applied to a dominant ecological or evolutionary viewpoint; for example, during earlier decades the dominant paradigm in *ecology* held that communities were shaped by equilibrial processes.

pelagic Relating to or living in a body of water.

phase state The condition of an ecosystem with regard to form, composition, or structure; it may be dominated by one organism, as in a sea urchin barrens.

phenology The scientific study of flowering, breeding, migration, and other periodic biological phenomena as they relate to climate.

phytoplankton A type of photosynthesizing plankton that makes up the lowest *trophic* level in aquatic systems and is fed on by *zooplankton*.

pinniped Member of an order of carnivorous marine mammals that have flippers and bodies that evolved for streamlined swimming; includes seals and walruses.

piscivores Fish or other animals, such as shorebirds, that eat fish.

planktivores Organisms, such as *zooplanton*, that eat other plankton, usually *phytoplankton*.

Pleistocene An epoch of the Quaternary period, also known as the Ice Age, extending from the end of the Pliocene, some 1.64 million years ago, to the beginning of the Holocene, approximately 10,000 years ago.

primary production The process of creating organic compounds from carbon dioxide through photosynthesis or chemosynthesis. All life relies on this process.

producers The organisms that form the lowest level of the food chain, that is, plants, also called *autotrophs*.

recruitment Survival of juveniles (plants or animals) for a sufficient period that they may reach adulthood.

recruitment gap Missing age classes in a tree community caused by chronic *herbivory*.

remote sensing Any technique for analyzing landscape patterns and trends using low-altitude aerial photography or satellite imagery. Any environmental measurement done from a distance.

rocky intertidal zone A narrow zone between the coast and the ocean revealed at low tide, characterized by breaking waves and an environment exposed to the air for part of the day (see *littoral*).

scale The magnitude of a region or process. Refers to both size, as with a relatively small-scale patch or a relatively large-scale landscape, and time, as with relatively rapid ecological *succession* or slow evolution of species.

secondary effects The situation in which population loss of a species affects other species, often through *trophic* interactions; also called indirect effects.

seral community An intermediate stage in ecological *succession* as a community advances to the climax stage.

sessile organisms Organisms that are unable to move, such as barnacles.

species diversity Usually synonymous with *species richness* but may also include the proportional distribution of species.

species richness The number of species in a region, site, or sample.

speciose Containing many species; high in overall *biodiversity*.

stochastic Random; particularly any random process, such as mortality, attributable to extreme weather, disease, or things beyond human control.

strongly interacting species Species that have a large effect on the other species with which they interact. Communities and ecosystems may have many strong interactors, occurring at all *trophic* levels. This general term can include *keystone species*.

succession The natural, sequential change of species composition in a community in a given area.

symbiosis A close association of two or more species.

sympatric Occupying the same or overlapping geographic areas without interbreeding.

taxon A taxonomic category or group, such as a phylum, order, family, genus, or species (plural form is *taxa*).

top-down control Regulation of lower *food web* components by an upper-level predator.

transect A line drawn through an area used to sample and monitor organisms or conditions. Typically it includes things that are a defined distance from both sides of the center line, thus creating a strip.

trophic Of or pertaining to nutrition, as in a trophic level of a *food web*.

ungulate A hoofed mammal.

watershed The land area that drains into a stream; the watershed for a major river may encompass a number of smaller watersheds that ultimately combine.

woody plants Plants that retain some living woody material at or above ground level through the nongrowing season; also called trees or shrubs. Cacti are exceptions to this, as they retain living material during the nongrowing season but are not trees or shrubs.

zooplankton Tiny, free-floating organisms, mainly crustaceans and fish larvae, characterized by involuntary movement. They cannot produce their own food and feed on *phytoplankton*.

Index

Note: Italicized page numbers indicate photographs or figures.